EVOLUTIONARY COMPUTATION

Books of Related Interest from IEEE Press . . .

EVOLUTIONARY COMPUTATION: The Fossil Record
Edited by David Fogel
1998   Hardcover   640 pp   IEEE Order No. PC5737   ISBN 0-7803-3481-7

FUZZY SYSTEMS DESIGN PRINCIPLES: Building Fuzzy If-Then Rule Bases
R. C. Berkan, and Sheldon Trubatch
1997   Hardcover   520 pp   IEEE Order No. PC5622   ISBN 0-7803-1151-5

EVOLUTIONARY COMPUTATION

Toward a New Philosophy
of Machine Intelligence

Second Edition

David B. Fogel
Natural Selection, Inc.

IEEE Neural Networks Council, Sponsor

**IEEE
PRESS**

The Institute of Electrical and Electronics Engineers, Inc., New York

This book may be purchased at a discount from the
publisher when ordered in bulk quantities. Contact:

IEEE Press Marketing
Attn: Special Sales
445 Hoes Lane, P.O. Box 1331
Piscataway, NJ 08855-1331
Fax: (732) 981-9334

For more information about IEEE Press products, visit the
IEEE Press Home Page: http://www.ieee.org/organizations/pubs/press

Printed in the United States of America

10   9   8   7   6   5   4   3   2   1

ISBN 0-7803-5379-X

IEEE Order Number: PC5818

5-23-00

Library of Congress Cataloging-in-Publication Data

Fogel, David B.
    Evolutionary computation : toward a new philosophy of machine
intelligence / David B. Fogel. — 2nd ed.
        p.   cm.
    "IEEE Neural Networks Council, sponsor."
    Includes bibliographical references.
    ISBN 0-7803-5379-X
    1. Computer simulation.   2. Artificial intelligence.
3. Evolutionary computation.   I. IEEE Neural Networks Council.
II. Title
QA76.9.C65F64   2000
006.3—dc21   99-21040
                                                    CIP

To
Wirt and Tony

CONTENTS

PREFACE TO THE
SECOND EDITION

Evolutionary computation is enjoying a renaissance. Conservatively, well over 2000 papers have been published in evolutionary computation since 1995 when the first edition of this book appeared and became a featured selection of the Library of Science Book Club. Publications on simulated evolution are now part of mainstream science and are no longer something out of the ordinary. Several evolutionary computation journals are now available, with some reaching thousands of readers worldwide (for example, the *IEEE Transactions on Evolutionary Computation*). The prospect of pursuing research in evolutionary computation has never been better.

Despite this visibility and acceptance, evolutionary computation and artificial intelligence (AI) still remain mostly disparate endeavors. The traditional perspective of programming human knowledge into a computer fact by fact, inference by inference, endures.

The crowning achievement of this approach was evidenced in May 1997, when the program Deep Blue defeated Garry Kasparov, the world chess champion. It was a triumph not only of human ingenuity, but also of the speed of computers for searching many moves ahead in the game, the design of application-specific hardware, and hand-tuned evaluation functions to accomplish by brute force what humans do with elegance. It was not, however, a triumph of AI. Indeed, the program that defeated Kasparov was not intelligent in any real sense of the word. The difference between Deep Blue and a calculator is only superficial: Beneath the facade is a preprogrammed "canned" procedure that learns nothing about the world in which it interacts. The headlines that read "Computer Beats Champion" were misleading. A more accurate account would have been "Humans Devise Tool to Beat Champion"—the tool simply happened to be a computer program. With respect to the intelligence of that program, it could just as well have been a hammer.

Deep Blue, like the majority of efforts in AI, simulates symptoms of intelligent behavior as observed in humans. The benefit of this approach is that it can generate highly optimized behavior, even to the extent that it surpasses human performance. The drawbacks are two-fold. The most obvious shortcoming is that this optimized behavior is applicable in only a very limited domain. The more fundamental flaw is that this approach does not contribute to an understanding of the intelligence that we observe in humans, animals, social groups, or even in the evolution of life itself.

Intelligence can be defined in terms of the capability of a system to adapt its behavior to meet its goals in a range of environments. The immediate question is then: How intelligent is Deep Blue? It plays an outstanding game of chess. Can it play checkers? Can it play tic-tac-toe? Can it discover new strategies for negotiating treaties when the parties involved have conflicting goals? The fact that these are rhetorical questions is evidence of how little has really been accomplished.

Genesereth and Nilsson (1987) suggested that the ultimate goal of AI is a theory that both explains the behavior of naturally occurring intelligent entities and indicates the manner in which artificial entities can be designed to yield intelligent behavior. The argument offered in this book is that the process of evolution accounts for such behavior and provides the foundation for the design of artificially intelligent machines.

In support of this thesis, I offer several examples. In each case, beginning with very restricted knowledge of the environment, an evolutionary algorithm learns how to solve the task at hand. This same algorithm can solve problems with a reasonable level of competency even when the conditions change. To make this point more evident, I have removed a previous example concerning the control of an unstable cart-pole system and have substituted new results of evolving strategies for playing checkers. The reader will note the relatively minor modifications that were required to transition from playing tic-tac-toe, as offered in the first edition and recapitulated here, to playing this new game. The versatility of the evolutionary procedure is one of its main strengths in serving as a basis for generating intelligent behavior, adapting to new challenges, and learning from experience.

This second edition also includes a significantly expanded fourth chapter concerning mathematical and empirical properties of evolutionary algorithms. There have been several important contributions to the theory of evolutionary computation since the publication of the first edition. Two of these have been offered by David Wolpert and William Macready. The first established that no evolutionary algorithm can be superior for all possible problems (i.e., the so-called "no free lunch" theorem), and the second identified a flaw in previous theory that served as the foundation for much work in one subset of evolutionary computation. This edition reviews these novel findings, together with other recent discoveries, so that the student of evolutionary computation can have a ready reference for the current state of knowledge in the field.

   This book is intended not only for students and practitioners of computer science and engineering, but also for the general reader who is curious about the prospects of creating intelligent machines. Its purpose will be served if it continues to advance interest, experimentation, and analysis in the growing field of evolutionary computation.

David B. Fogel
Natural Selection, Inc.

REFERENCE

Genesereth, M. R., and N. J. Nilsson (1987). *Logical Foundations of Artificial Intelligence.* Los Altos, CA: Morgan Kaufmann.

PREFACE TO THE
FIRST EDITION

Intelligence and evolution are intimately connected. Intelligence is a natural part of life. It is also, however, a mechanical process that can be simulated and emulated. Intelligence is not a property that can be limited philosophically solely to biological structures. It must be equally applicable to machines. Although such efforts for generating intelligence in machines are typically described by the term *artificial intelligence*, this is in fact a misnomer. If intelligence were to be properly represented as to its process there would be nothing artificial about it. If the process is understood, methods for its generation should converge functionally and become fundamentally identical, relying on the same physics whether the intelligence occurs in a living system or a machine.

The majority of research in artificial intelligence has simulated symptoms of intelligent behavior as observed in humans. In contrast, I propose a definition of intelligence that does not rely only in its comparisons to human behavior. Intelligence is defined as the capability of a system to adapt its behavior to meet its goals in a range of environments. The form of the intelligent system is irrelevant, for its functionality is the same whether intelligence occurs within an evolving species, an individual, or a social group. Rather than focus on ourselves and attempt to emulate our own behaviors and cognitive processes, it would appear more prudent to recognize that we are products of evolution and that by modeling evolutionary processes we can create entities capable of generating intelligent behavior. Evolutionary computation, the field of simulating evolution on a computer, provides the basis for moving toward a new philosophy of machine intelligence.

Evolution is categorized by several levels of hierarchy: the gene, the chromosome, the individual, the species, the ecosystem. Thus there is an inevitable choice that must be made when constructing a simulation of evolution. Inevitably, attention must be focused at a particular level in the hierarchy, and the remainder of the simulation is to some extent determined by that perspective. Ultimately, the question that must be answered is, "What exactly is being evolved?" All the symptoms of the simulation will reflect the answer to this question, whether by conscious design or not. If intelligence is viewed in terms of adaptation, then the answer to the question must be that what is evolved is functional behavior. To construct a useful model, evolution must be abstracted in terms of the behavioral relationships between units of evolution, rather than the mechanisms that give rise to these relationships.

The result of such modeling is a series of optimization algorithms that rely on very simple rules. These various procedures are implemented as population-based searches over a fitness response surface. The optimization process inherent in selection iteratively improves the quality of these solutions. The rate of improvement is primarily determined by the shape of the response surface, but the procedures generally converge to near-optimal solutions despite the existence of topological pathologies. In some cases, it is provable that the procedures will asymptotically converge to the best solutions; in others, it is provable that they will never converge at all. Nonetheless, these methods offer potential for addressing engineering problems that have resisted solution by classic techniques.

More importantly, however, these methods may be used to address a range of problems, rather than any one specific problem. They have proven themselves to be robust and may be applied toward general problem solving. This latter attribute represents the greatest potential for evolutionary computation, yet there have been few investigations in this regard. The real promise of evolutionary computation remains mostly unfulfilled.

This book is an attempt to integrate the inspiration, philosophy, history, mathematics, actualizations, and perspectives of evolutionary computation. It will have been successful if this integration adds to the reader's understanding of the field of research. My hope is that it will encourage further investigation into the potential of these techniques for generating machine intelligence.

There are many people to thank for their time and effort spent helping me prepare the manuscript and offering comments on this work. Dudley Kay, Lisa Mizrahi, Val Zaborski, and Karen Henley of IEEE Press and copyeditor Kathleen Lafferty were most supportive. Teresa Moniz and Mike MacVittie assisted with the artwork, and I would like to thank them and ORINCON Corporation, and especially Terry Rickard, for making their time and skills available. I also greatly appreciate the help received from Scott Haffner, Brian Moon, and Rob Redfield converting and debugging computer programs for execution on different machines. Chapters and sections, including those taken from recent publications, were reviewed by Lee Altenberg, Russell Anderson, Pete Angeline, Valerie Atmar, Wirt Atmar, Thomas Bäck, Rick Barton, Hans Bremermann, Tom Brotherton, Cliff Brunk, George Burgin, Michael Conrad, Lawrence (Dave) Davis, Hugo de Garis, Ken De Jong, Marco Dorigo, Tom

English, Gary Fogel, Larry Fogel, Steve Frank, John Grefenstette, Paul Harrald, David Honig, Gerry Joyce, Mark Kent, Avi Krieger, Charles Marshall, Ernst Mayr, Zbyszek Michalewicz, Mike O'Hagan, Bill Porto, Bob Powell, Tom Ray, Ingo Rechenberg, Bob Reynolds, Matt Rizki, Stuart Rubin, Günter Rudolph, N. (Saravan) Saravanan, Bill Schopf, Hans-Paul Schwefel, Brad Scurlock, Tony Sebald, Pat Simpson, Chris Wills, Xin Yao, and several anonymous referees. Their criticisms have been invaluable. I would be remiss if I did not acknowledge the encouragement of John McDonnell, Ward Page, and Jennifer Schlenzig, and I am deeply grateful to Wirt Atmar and Tony Sebald for years of guidance and friendship. Finally, I thank my family for their constant support.

David B. Fogel
Natural Selection, Inc.

ACKNOWLEDGMENTS

I would like to thank IEEE Press for encouraging a second edition of this work, and in particular I am grateful to John Griffin, Marilyn Catis, Ken Moore, Mark Morrell, and Barb Soifer for their support. I would also like to thank the six technical reviewers for this edition, as well as Pete Angeline, Gary Fogel, Larry Fogel, and Bill Porto, for their advice and criticism.

David B. Fogel
Natural Selection, Inc.

CHAPTER 1

DEFINING ARTIFICIAL INTELLIGENCE

1.1 BACKGROUND

Although most scientific disciplines, such as mathematics physics, chemistry, and biology, are well defined, the field of artificial intelligence (AI) remains enigmatic. Indeed, Hofstadter (1985, p. 633) remarks, "The central problem of AI is the question: *What is the letter 'a'?* Donald Knuth, on hearing me make this claim once, appended, 'And what is the letter "i"?'—an amendment that I gladly accept." Despite nearly 50 years of research in the field, no widely accepted definition of artificial intelligence exists.

Artificial intelligence is sometimes defined as the manipulation of symbols for problem solving (e.g., Buchanan and Shortliffe, 1985, p. 3) or by tautological statements such as artificial intelligence is "the part of computer science concerned with developing intelligent computer programs" (Waterman, 1986, p. 10). Rich (1983, p. 1) offered, "Artificial intelligence (AI) is the study of how to make computers do things at which, at the moment, people are better." But this definition, if regarded statically, precludes the very existence of artificial intelligence. Once a computer program exceeds the capabilities of a human, the program is no longer in the domain of AI.

The majority of proffered definitions of artificial intelligence rely on comparisons to human behavior. Staugaard (1987, p. 23) attributes a definition to Marvin Minsky—"the science of making machines do things that would require intelligence if done by men"—and suggests that some define AI as the "mechanization, or duplication, of the human thought process." Charniak and McDermott (1985, p. 6) offer, "Artificial intelligence is the study of mental faculties through the use of computational models," while Schildt (1987, p. 11) claims, "An intelligent program is one that exhibits behavior similar to that of a human when confronted with a similar problem. It is not necessary that the program actually solve, or attempt to solve, the problem in the same way that a human would."

The question, "What is AI?" would become mere semantics if only the answers did not suggest or imply radically different avenues of research: "Some researchers simply want machines to do the various sorts of things that people call intelligent. Others hope to understand what enables people to do such things. Still other researchers want to simplify programming" (Minsky, 1991). Yet Minsky also laments, "Why can't we build, once and for all, machines that grow and improve themselves by learning from experience? Why can't we simply explain what we want, and then let our machines do experiments or read some books or go to school, the sorts of things that people do. Our machines today do no such things."

1.2   THE TURING TEST

Turing (1950) considered the question, "Can machines think?" Rather than define the terms *machines* or *think,* he proposed a test which begins with three people, a man (A), a woman (B), and an interrogator (C). The interrogator is to be separated from both A and B, say, in a closed room (Figure 1-1) but may

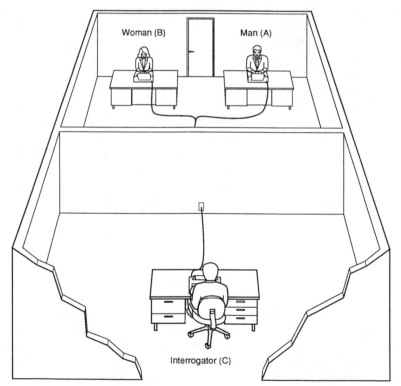

Figure 1-1 The Turing test. An interrogator (C) questions both a man (A) and a woman (B) and
attempts to determine which is the woman.

ask questions of both A and B. The interrogator's objective is to determine which of A and B is the man and which is the woman. It is A's objective to cause C to make an incorrect identification. Turing (1950) provided the following example of a question posed to the man:

"C: Will X [C's name for A] please tell me the length of his or her hair?"

"A: My hair is shingled, and the longest strands are about nine inches long."

Player A may be deceitful, if he so desires. In contrast, the object for B is to help the interrogator. Turing suggested that the probable best strategy for her is to give truthful answers. In order that the pitch of the voice or other external clues may not aid in C's decision, a teleprinter was to be used for communication between the rooms.

Turing then replaced the original question, "Can machines think?" with the following: "We now ask the question, 'What will happen when a machine takes the part of A in this game?' Will the interrogator decide wrongly as often when the game is played like this as he does when the game is played between a man and a woman?" (Turing, 1950). This question separates the physical and intellectual capacities of humans. The form of interrogation prevents C from using sensory information regarding A's or B's physical characteristics. Presumably, if the interrogator were able to show no increased ability to decide between A and B when the machine was playing as opposed to when the man was playing, then the machine would be declared to have passed the test. Whether or not the machine should then be judged capable of thinking was left unanswered. Turing (1950) in fact dismissed the original question as being "too meaningless to deserve discussion."

Turing (1950) limited the possible machines to be the set of all digital computers. He indicated through considerable analysis that these machines are *universal,* that is, all computable processes can be executed by such machines. With respect to the suitability of the test itself, Turing (1950) thought the game might be weighted "too heavily against the machine. If the man were to try and pretend to be the machine he would clearly make a very poor showing." (Hofstadter, 1985, pp. 514–520, relates an amusing counterexample in which he was temporarily fooled in such a manner.)

Turing (1950) considered and rejected a number of objections to the plausibility of a "thinking machine," although somewhat remarkably he felt an argument supporting the existence of extrasensory perception in humans was the most compelling of all objections. The *Lady Lovelace* objection (Countess of Lovelace, 1842), referring to a memoir by the Countess of Lovelace on Babbage's Analytical Engine, is the most common of present refutations of a thinking machine. The argument asserts that a computer can only do what it is programmed to do and therefore will never be capable of generating anything new. Turing (1950) countered this argument by equating it with a statement that a machine can never take us by surprise. But he noted that machines often act in unexpected ways because the entire determining set of initial conditions of the machine is generally unknown; therefore, an accurate prediction of all possible behavior of the mechanism is impossible.

Moreover, he suggested that a thinking machine should be a learning machine, capable of altering its own configuration through a series of rewards and punishments. Thus it could modify its own programming and generate unexpected behavior. He speculated that "in about fifty years' time it will be possible to programme computers, with a storage capacity of about 10^9 [bits], to make them play the imitation game so well that an average interrogator will not have more than a 70 per cent chance of making the right identification after five minutes of questioning" (Turing, 1950).

1.3   SIMULATION OF HUMAN EXPERTISE

The acceptance of the "Turing Test" focused attention on mimicking human behavior. At the time (1950), it was beyond any reasonable consideration that a computer could pass the Turing Test. Rather than focus on imitating human behavior in conversation, attention was turned to more limited domains of interest. Simple two-person games of strategy were selected. These games received attention for at least three reasons: (1) Their rules are static, known to both players, and easy to express in a computer program, (2) they are examples from a class of problems concerned with reasoning about actions, and (3) the ability of a game-playing computer can be measured against human experts.

The majority of research in game playing has been aimed at the development of heuristics that can be applied to two-person, zero-sum, nonrandom games of perfect information (Jackson, 1985). The term *zero-sum* indicates that any potential gain to one player will be reflected as a corresponding loss to the other player. The term *nonrandom* means that the allocation and positioning of resources in the game (e.g., pieces on a chess board) is purely deterministic. *Perfect information* indicates that both players have complete knowledge regarding the disposition of both players' resources (e.g., tic-tac-toe, not poker).

The general protocol was to examine an expert's decisions during a game so as to discover a consistent set of parameters or questions that are evaluated during his or her decision-making process. These conditions could then be formulated in an algorithm capable of generating behavior similar to that of the expert when faced with identical situations. It was believed that if a sufficient quantity or "coverage" of heuristics could be programmed into the computer, the sheer speed and infallible computational ability of the computer would enable it to match or even exceed the ability of the human expert.

1.3.1   Samuel's Checker Program

One of the earliest efforts along these lines was offered by Samuel (1959), who wrote a computer program that learned to play checkers. Checkers was chosen for several reasons: (1) There is no known algorithm that provides for a guaranteed win or draw, (2) the game is well defined with an obvious goal, (3) the rules are fixed and well known, (4) there are human experts who can be consulted and against which progress of the program can be tested, and (5) the activity is familiar to many people. The general procedure of the program was to look ahead a few moves at a time and evaluate the resulting board positions.

This evaluation was made with respect to several selected parameters. These parameters were then included in a linear polynomial with variable coefficients. The result of the polynomial indicated the worth of the prospective board under evaluation. The most critical and obvious parameter was the inability for one side or the other to move. This can occur only once in a game and was tested separately. Another clearly important consideration was the relative piece advantage. Kings were given 50 percent more weight than regular pieces. Samuel (1959) tried two alternative methods for including additional parameters. Initially, Samuel himself chose these terms, but he later allowed the program to make a subset selection from a large list of possible parameters.

To determine a move, the game tree of possible new boards was searched. A minimax procedure was used to discover the best move. The *ply,* or number of levels to be searched in the tree, was initially set at three unless the next move was a jump, the last move was a jump, or an exchange offer was possible. The analysis proceeded backward from the evaluated board position through the tree of possible moves, with the assumption that at each move the opponent would always attempt to minimize the machine's score while the machine would act to maximize its score. Under these conditions the search was continued until these circumstances were no longer encountered or until a maximum of 20 levels had been searched.

After initial experiments in which the selected polynomial had four terms (piece advantage, denial of occupancy, mobility, and a hybrid term that combined control of the center and piece advancement), the program was allowed to select a subset of 16 parameters from a list of 38 chosen parameters. Samuel allowed the computer to compete against itself; one version, Alpha, constantly modified the coefficients and parameters of its polynomial, and the other version, Beta, remained fixed (i.e., it was replaced by Alpha after a loss). A record of the correlation existing between the signs of the individual term contributions in the initial scoring polynomial and the sign of the change between the scores was maintained, along with the number of times that each particular term with a nonzero value was used. The coefficient for the polynomial term of Alpha with the then-largest correlation coefficient was set at a prescribed maximum value with proportionate values determined for all the remaining coefficients. Samuel noted some possible instabilities with this modification technique and developed heuristic solutions to overcome these problems. Term replacement was made when a particular parameter had the lowest correlation eight times. Upon reaching this arbitrary limit, it was placed at the bottom of the reserve list and the first parameter in the reserve list was inserted into the scoring polynomial.

After a series of 28 games, Samuel described the program as being a better-than-average player. "A detailed analysis of these results indicated that the learning procedure did work and that the rate of learning was surprisingly high, but that the learning was quite erratic and none too stable" (Samuel, 1959). Refinements were made to the method of altering the scoring polynomial to help prevent this problem.

In 1962, at the request of Edward Feigenbaum and Julian Feldman, Samuel arranged for a match between his program and Robert W. Nealy, a purported former Connecticut checkers champion. Samuel's program defeated Nealy, who commented (cited in Samuel, 1963):

> Our game . . . did have its points. Up to the 31st move, all of our play had been previously published, except where I evaded "the book" several times in a vain effort to throw the computer's timing off. At the 32–27 [a specific move] loser and onwards, all the play is original with us, so far as I have been able to find. It is very interesting to me to note that computer had to make several star moves in order to get the win, and that I had several opportunities to draw otherwise. That is why I kept the game going. The machine, therefore, played a perfect ending without one misstep. In the matter of the end game, I have not had such competition from any human being since 1954, when I lost my last game.

The moves of the game appear in Samuel (1963).

In retrospect, perhaps more acclaim was given to this result than was deserved. Schaeffer (1996, p. 94) indicated that Nealy was in fact not a former Connecticut state champion at the time of the match against Samuel's program, although he did earn that title in 1966, four years later. Moreover, Nealy did not enter the U.S. Championship checkers tournament, and thus the strength of his play at the national level was based more on opinion than on record. Schaeffer (1996, pp. 94–95) reviewed the sequence of moves from the Nealy match, and with the aid of *Chinook* (a current world champion artificial intelligence checkers program designed by Schaeffer and his colleagues), indicated that Nealy made several blunders during the game and that Samuel's checkers program also did not capitalize on possible opportunities. In sum, the glowing description that Nealy gave of Samuel's program's endgame is well accepted in the literature but is an overstatement of the program's ability. Schaeffer (1996, p. 97) also reported that, in 1966, Samuel's program was played against the two persons vying for the world championship. Four games were played against each opponent, with Samuel's program losing all eight matches.

1.3.2   Chess Programs

Researchers in artificial intelligence have also been concerned with developing chess programs. Initial considerations of making machines play chess date to Charles Babbage (1792–1871). Babbage had described the Analytical Engine, a theoretic mechanical device that was a digital computer, although not electronic. This machine was never built (but an earlier design, the Difference Engine, was in fact successfully constructed only as recently as 1991; Swade, 1993). Babbage recognized that, in principle, his Analytical Engine was capable of playing games such as checkers and chess by looking forward to possible alternative outcomes based on current potential moves.

Shannon (1950) was one of the first researchers to propose a computer program to play chess. He, like Samuel later, chose to have an evaluation function such that a program could assess the relative worth of different configurations of pieces on the board. The notion of an evaluation function has been an integral component of every chess program ever since. The suggested parameters included material advantage, pawn formation, positions of pieces, mobility, commitments, attacks, and options (cited in Levy and Newborn, 1991, pp. 27–28). Shannon noted that the best move can be found in at least two ways, although the methods may be combined: (1) Search to a given number of moves ahead and then use a minimax algorithm or (2) selectively search different branches of the game tree to different levels (i.e., moves ahead). The second method offers the advantage of preventing the machine from wasting time searching down branches in which one or more bad moves must be made. This method, later termed the *alpha-beta algorithm,* has been incorporated in many current chess playing programs.

Turing (1953) is credited with writing the first algorithm for automatic chess play. He never completed programming the procedure on a computer but was able to play at least one game by hand simulation. Turing's evaluation function included parameters of mobility, piece safety, castling, pawn position, and checks and mate threats. The one recorded game (cited in Levy and Newborn, 1991, pp. 35–38) used a search depth of two ply and then continued the search down prospective branches until "dead" positions (e.g., mate or the capture of an undefended piece) were reached. In this game, the algorithm was played against "a weak human opponent" (Levy and Newborn, 1991, p. 35) and subsequently lost. Turing attributed the weakness of the program to its "caricature of his own play" (cited in Levy and Newborn, 1991, p. 38).

The first documented working chess program was created in 1956 at Los Alamos. An unconfirmed account of a running program in the Soviet Union was reported earlier by *Pravda* (Levy and Newborn, 1991, p. 39). Bernstein et al. (1958) described their computer program, which played a fair opening game but weak middle game because the program only searched to a depth of four ply. Newell et al. (1958) were the first to use the alpha-beta algorithm (Shannon, 1950). Greenblatt et al. (1967) are credited with creating the first program, called MACHACK VI, to beat a human in tournament play. The program was made an honorary member of the United States Chess Federation (USCF), receiving their rating of 1640. MACHACK VI used a search of at least nine ply.

In 1978, CHESS 4.7, a revised version of a program originally written by Atkin, Gorlen, and Slate of Northwestern University, defeated David Levy, Scottish chess champion, in a tournament game. Levy was "attempting to beat the program at its own game" and returned in the next match to a "no-nonsense approach," presumably to win (Levy and Newborn, 1991, pp. 98, 100). BELLE, written by Thompson and Condon, was the first program that qualified, in 1983, for the title of U.S. Master.

In the 1980s, efforts were directed at making application-specific hardware capable of searching large numbers of possible boards and quickly calculating appropriate evaluations. Berliner created HITECH, a 64-processor system. Hsu produced an even more powerful chip and its resident program, now known as DEEP THOUGHT, quickly outperformed HITECH. DEEP THOUGHT was able to search to a level of 10 ply and became the first program to defeat a world-class grand master, Bent Larsen. In 1989, DEEP THOUGHT, then rated at 2745, played a four-game match against David Levy. Levy admits, "It was the first time that [I] had ever played a program rated higher than [I] was at [my] best" (Levy and Newborn, 1991, p. 127), and correctly predicted that the machine would win 4–0. In 1990, Anatoly Karpov, the former world champion, lost a game to a MEPHISTO chess computer while giving a simultaneous exhibition against 24 opponents.

The pinnacle of beating a human world champion in match play finally was achieved in May 1997 when IBM's Deep Blue, the successor to DEEP THOUGHT, defeated Garry Kasparov, scoring two wins, one loss, and three draws. The previous year, Kasparov had defeated Deep Blue, scoring three wins, one loss, and two draws. The computer horsepower behind Deep Blue included 32 parallel processors and 512 custom chess ASICs which allowed a search of 200 million chess positions per second (Hoan, cited in Clark, 1997). Although the event received wide media attention and speculation that computers had become "smarter than humans," surprisingly little attention was given to the event in scientific literature. McCarthy (1997) offered that Deep Blue was really "a measure of our limited understanding of the principle of artificial intelligence (AI) . . . . this level of play requires many millions of times as much computing as a human chess player does." Indeed, there was no automatic learning involved in Deep Blue. A. Joseph Hoan Jr., a member of the team that developed Deep Blue, remarked (in Clark, 1997): "we spent the whole year with chess grand master, Joel Benjamin, basically letting him beat up Deep Blue—making it make mistakes and fixing all those mistakes. That process may sound a little clunky, but we never found a good way to make automatic tuning work." Between games, adjustments were made to Deep Blue based on Kasparov's play, but these again were made by the humans who developed Deep Blue, not by the program itself.

Judging by the nearly linear improvement in the USCF rating of chess programs since the 1960s (Levy and Newborn, 1991, p. 6), the efforts of researchers to program computers to play chess must be regarded as highly successful. But there is a legitimate question as to whether or not these programs are rightly described as intelligent. Schank (1984, p. 30) commented, "The moment people succeeded in writing good chess programs, they began to wonder whether or not they had really created a piece of Artificial Intelligence. The programs played chess well because they could make complex calculations with extraordinary speed, not because they knew the kinds of things that human chess masters know about chess." Simply making machines do things that people would describe as requiring intelligence is insufficient (cf. Staugaard, 1987, p. 23). "Such programs did not embody intelligence and did not contribute to the quest for intelligent machines. A person isn't intelligent because he or she is a chess master; rather, that person is able to master the game of chess because he or she is intelligent" (Schank, 1984, p. 30).

1.3.3 Expert Systems

The focus of artificial intelligence narrowed considerably from the early 1960s through the mid-1980s (Waterman, 1986, p. 4). Initially, the desire was to create general problem-solving programs (Newell and Simon, 1963), but when preliminary attempts were unsuccessful, attention was turned to the discovery of efficient search mechanisms that could process complex data structures. The focus grew even more myopic in that research was aimed at applying these specific search algorithms (formerly termed *heuristic programming*) to very narrowly defined problems. Human experts were interrogated about their knowledge in their particular field of expertise, and this knowledge was then represented in a form that supported reasoning activities on a computer. Such an *expert system* could offer potential advantages over human expertise: It is "permanent, consistent, easy to transfer and document, and cheaper" (Waterman, 1986, p. xvii). Nor does it suffer from human frailties such as aging, sickness, or fatigue.

The programming languages often used in these applications are LISP (McCarthy et al., 1962) and Prolog (invented by Colmerauer, 1972, as cited in Covington et al., 1988, p. 2). To answer questions that are posed to the system, an *inference engine* (a program) is used to search a knowledge base.

Knowledge is most frequently represented using first-order predicate calculus, production rules, semantic networks, and frames. For example, a knowledge base might include the facts: Larry is a parent of Gary and David. This might be represented in Prolog as

1. parent(larry, gary), and
2. parent(larry, david).

(Versions of Prolog often reserve the use of capitals for variables.) If one were to be able to interrogate about whether or not a person was a child of Larry, the additional facts

1. child(gary, larry), and
2. child(david, larry),

or the rule

1. child(X, Y) :-
       parent(Y, X).

could be included (:- denotes "if"). The computer has no intrinsic understanding of the relationships "parent" or "child." It simply has codings (termed *functors*) that relate "gary," "david," and "larry."

With the existence of such a knowledge base, it becomes possible to query the system about the relationship between two people. For example, if one wanted to know whether or not "david" was the parent of "gary," one could enter

? - parent(david, gary).

The inference engine would then search the knowledge base of rules and facts and fail to validate "parent(david, gary)" and therefore would reply, "no." More general questions could be asked, such as

? - parent(larry, X).

where X is a variable. The inference engine would then search the knowledge base, attempting to match the variable to any name it could find (in this case, either "gary" or "david").

Although these examples are extremely simple, it is not difficult to imagine more complex relationships programmed in a knowledge base. The elements in the knowledge base need not be facts, but may be conjectures with degrees of confidence assigned by the human expert. The knowledge base may contain conditional statements (production rules), such as, "IF *premise* THEN *conclusion*" or "IF *condition* WITH *certainty greater than x* THEN *action*." Through the successive inclusion of broad-ranging truths to very specific knowledge about a limited domain, a versatile knowledge base and query system can be created.

DENDRAL, a chemistry program that processed mass spectral and nuclear magnetic response data to provide information regarding the molecular structure of unknown compounds, was one of the first such systems. The program was started in the mid-1960s and was subsequently refined and extended by several researchers (e.g., Feigenbaum et al., 1971; Lindsay et al., 1980). MYCIN (Shortliffe, 1974; Adams, 1976; Buchanan and Shortliffe, 1985), a program to diagnose bacterial infections in hospital patients, was an outgrowth of the "knowledge-based" DENDRAL project. Other examples of well-known knowledge-based systems can be found in Bennett and Hollander (1981), Barr and Feigenbaum (1981), and Lenat (1983).

The expert system PROSPECTOR was developed by the Stanford Research Institute to aid exploration geologists in the search for ore deposits (Duda et al., 1978). Work on the system continued until 1983 (Waterman, 1986, p. 49). Nine different mineral experts contributed to the database; it contains over 1,000 rules and a taxonomy of geological terms with more than 1,000 entries. The following sequence represents an example of PROSPECTOR receiving information from a geologist:

"1: THERE ARE DIKES
     (Dike) (5)
 2: THERE ARE CRETACEOUS DIORITES
     (Cretaceous diorites) (5)
 3: THERE IS PROBABLY SYENODIORITE
     (Monzonite) (3)
 4: THERE MIGHT BE SOME QUARTZ MONZONITE
     (Quartz-monzonite) (2)"

(Waterman, 1986, p. 51). The values in parentheses represent the degree of certainty associated with each statement (-5 indicates complete certainty of absence while $+5$ indicates complete certainty of existence). The nouns in parentheses represent the internally stored name for the substance described. Through subsequent questioning of the human expert by the expert system, the program is able to offer a conjecture such as

My certainty in (PCDA) [Type-A porphyry copper deposit] is now: 1.683

followed by a detailed listing of the rules and facts that were used to come to this conclusion.

In 1980, PROSPECTOR was used to analyze a test drilling site near Mount Tolman in eastern Washington that had been partially explored. PROSPECTOR processed information regarding the geological, geophysical, and geochemical data describing the region and predicted the existence of molybdenum in a particular location (Campbell et al., 1982). "Subsequent drilling by a mining company confirmed the prediction as to where ore-grade molybdenum mineralization would be found and where it would not be found" (Waterman, 1986, p. 58).

Waterman (1986, pp. 162–199) described the construction of an expert system and discussed some potential problems. For example, the number of rules required for a given application may grow very large. "PUFF, an expert system that interprets data from pulmonary function tests, had to have its number of rules increased from 100 to 400 just to get a 10 percent increase in performance" (Waterman, 1986, p. 182). PUFF required five person-years to construct. There are no general techniques for assessing a system's completeness or consistency. Nevertheless, despite the procedural difficulties associated with constructing expert systems, useful programs have been created to address a wide range of problems in various domains including medicine, law, agriculture, military sciences, geology, and others (Waterman, 1986, p. 205). Building expert systems has now become routine (Barr et al., 1989, p. 181).

1.3.4   A Criticism of the Expert Systems or Knowledge-Based Approach

There is some question whether or not the research in expert or knowledge-based systems truly advances the field of artificial intelligence. Dreyfus and Dreyfus (1984, 1986) claimed that when a beginner is first introduced to a new task, such as driving a car, he or she is taught specific rules to follow (e.g., maintain a two-second separation between yourself and the car in front of you). But as the beginner gains experience, less objective cues are used. "He listens to the engine as well as looks at his speedometer for cues about when to shift gears. He observes the demeanor as well as the position and speed of pedestrians to anticipate their behavior. And he learns to distinguish a distracted or drunk driver from an alert one" (Dreyfus and Dreyfus, 1984). It is difficult to believe that the now-expert driver is relying on rules in making these classifications. "Engine sounds cannot be adequately captured by words, and no list of facts about a pedestrian at a crosswalk can enable a driver to predict his behavior as well as can the experience of observing people crossing streets under a variety of conditions" (Dreyfus and Dreyfus, 1984). They asserted that when a human expert is interrogated by a "knowledge-engineer" to assimilate rules for an expert system, the expert is "forced to regress to the level of a beginner and recite rules he no longer uses . . . there is no reason to believe that a heuristically programmed computer accurately replicates human thinking" (Dreyfus and Dreyfus, 1984).

Other problems occur in the design of expert systems. Human experts, when forced to verbalize rules for their behavior, may not offer a consistent set of explanations. There may be inherent contradictions in their rules. In addition, different experts will differ on the rules that should be employed. The question of how to handle these inconsistencies remains open and is often handled in an ad hoc manner by knowledge-engineers who do not have expertise in the field of application. Simply finding an expert can be troublesome, for most often there is no objective measure of "expertness." And even when an expert is found, there is always the chance that the expert will simply be wrong. History is replete with incorrect expertise (e.g., a geocentric solar system, and see Cerf and Navasky, 1984). Moreover, expert systems often generate preprogrammed behavior. Such behavior can be *brittle,* in the sense that it is well optimized for its specific environment, but incapable of adapting to any changes in the environment.

Consider the hunting wasp, *Sphex flavipennis.* When the female wasp must lay its eggs, its builds a burrow and hunts out a cricket, which it paralyzes with three injections of venom (Gould and Gould, 1985). The wasp then drags the cricket into the burrow, lays the eggs next to the cricket, seals the burrow, and flies away. When the eggs hatch, the grubs feed on the paralyzed cricket. Initially, this behavior appears sophisticated, logical, and thoughtful (Wooldridge, 1968, p. 70). But upon further examination, limitations of the wasp's behavior can be demonstrated. Before entering the burrow with the cricket, the wasp carefully positions its paralyzed prey with its antennae just touching the opening of the burrow and then scoots inside, 'inspects' its quarters, emerges, and drags the captive inside" (Gould and Gould, 1985). As noted by French naturalist Jean Henri Fabré, if the cricket is moved just slightly while the wasp is busy in its burrow, upon emerging the wasp will replace the cricket at the entrance and again go inside to inspect the burrow. "No matter how many times Fabré moved the cricket, and no matter how slightly, the wasp would never break out of the pattern. No amount of experience could teach it that its behavior was wrongheaded: its genetic inheritance had left it incapable of learning that lesson" (Gould and Gould, 1985).

The wasp's instinctive program is essentially a rule-based system that is crucial in propagating the species, unless the weather happens to be a bit breezy. "The insect, which astounds us, which terrifies us with its extraordinary intelligence, surprises us, the next moment, with its stupidity, when confronted with some simple fact that happens to lie outside its ordinary practice" (Fabré, cited in Gould and Gould, 1985). Genetically hard-coded behavior is inherently brittle. Similarly, an expert system chess program might do very well, but if the rules of the game were changed, even slightly, by perhaps allowing the king to move up to two squares at a time, the expert system might no longer be expert. This brittleness of domain-specific programs has been recognized for many years (Samuel, 1959).

1.3.5 Fuzzy Systems

Another procedural problem associated with the construction of a knowledge base is that when humans describe complex environments, they do not typically speak in absolutes. Linguistic descriptors of real-world circumstances are not precise but rather are "fuzzy." For example, when one describes the optimum behavior of an investor interested in making money in the stock market, the adage is "buy low, sell high." But how low is "low"? And how high is "high"? It is unreasonable to suggest that if the price of the stock climbs to a certain precise value in dollars per share, then it is high; yet if it were only $0.01 lower, it would not be high. Useful descriptions need not be of a binary or crisp nature.

Zadeh (1965) introduced the notion of "fuzzy sets." Rather than describing elements as being either in a given set or not, membership in the set was viewed as a matter of degree ranging over the interval $[0, 1]$. A membership of 0.0 indicates that the element absolutely is not a member of the set, and a membership of 1.0 indicates that the element absolutely is a member of the set. Intermediate values indicate degrees of membership. The choice of the appropriate membership function to describe elements of a set is left to the researcher.

Negoita and Ralescu (1987, p. 79) note that descriptive phrases such as "numbers approximately equal to 10" and "young children" are not tractable by methods of classic set theory or probability theory. There is an undecidability about the membership or nonmembership in a collection of such objects, and there is nothing random about the concepts in question. A classic set can be represented precisely as a binary valued function $f_A: X \rightarrow \{0, 1\}$, the *characteristic function,* defined as

$$f_A(x) = \begin{cases} 1, \text{if } x \in A; \\ 0, \text{otherwise.} \end{cases}$$

The collection of all subsets of X (the power set of X) is denoted

$$P(X) = \{A | A \text{ is a subset of } X\}.$$

In contrast, a fuzzy subset of X is represented by a *membership function:*

$$u : X \rightarrow [0, 1].$$

The collection of all fuzzy subsets of X (the fuzzy power set) is denoted by $F(X)$:

$$F(X) = \{u | u : X \rightarrow [0, 1]\}.$$

It is natural to enquire as to the effect of operations such as union and intersection on such fuzzy sets. If u and v are fuzzy sets, then

$$(u \text{ or } v)(x) = \max[u(x), v(x)]$$
$$(u \text{ and } v)(x) = \min[u(x), v(x)].$$

Other forms of these operators have been developed (Yager, 1980; Dubois and Prade, 1982; Kandel, 1986, pp. 143–149). One form of the complement of a fuzzy set, $u : X \rightarrow [0, 1]$ is

$$u^c(x) = 1 - u(x).$$

Other properties of fuzzy set operations, such as commutativity, associativity, and distributivity, as well as other operators such as addition, multiplication, and so forth, may be found in Negoita and Ralescu (1987, pp. 81–93).

It is not difficult to imagine a *fuzzy system* that relates fuzzy sets in much the same manner as a knowledge-based system. The range of implementation and reasoning methodologies is much richer in fuzzy logic. The rules are simply fuzzy rules describing memberships in given sets rather than absolutes. Such systems have been constructed and Bezdek and Pal (1992) give a comprehensive review of efforts in fuzzy systems from 1965.

1.3.6 Perspective on Methods Employing Specific Heuristics

Human experts can only give dispositions or rules (fuzzy or precise) for problems in their domain of expertise. There is a potential difficulty when such a system is required to address problems for which there are no human experts. Schank (1984, p. 34) states, definitively, "Expert systems are horribly misnamed, since there is very little about them that is expert . . . while potentially useful, [they] are not a theoretical advance in our goal of creating an intelligent machine." He continues:

> The central problem in AI has little to do with expert systems, faster programs, or bigger memories. The ultimate AI breakthrough would be the creation of a machine that can learn or otherwise change as a result of its own experiences . . . like most AI terms, the words "expert system" are loaded with a great deal more implied intelligence than is warranted by their actual level of sophistication. . . . Expert systems are not innovative in the way the real experts are; nor can they reflect on their own decision-making processes.

This generalization may be too broad. Certainly, both expert and fuzzy systems can be made very flexible. They can generate new functional rules that were not explicitly stated in the original knowledge base. They can be programmed to ask for more information from human experts if they are unable to reach any definite (or suitably fuzzy) conclusion in the face of current information. Yet one may legitimately question whether the observed "intelligence" of the system should really be attributed to the system, or merely to the programmer who implemented knowledge into a fixed program. Philosophically, there appears to be little difference between such a hard-wired system and a simple calculator. Neither is intrinsically intelligent.

The widespread acceptance of the Turing Test has both focused and constrained research in artificial intelligence in two regards: (1) the imitation of human behavior and (2) the evaluation of artificial intelligence solely on the basis of behavioral response. But, "Ideally, the test of an effective understanding system is not the realism of the output it produces, but rather the validity of the method by which that output is produced" (Schank, 1984, p. 53).

Hofstadter (1985, p. 525) admitted to being an "unabashed pusher of the validity of the Turing Test as a way of operationally defining what it would be for
a machine to genuinely think." But he also cogently wrote while playing devil's
advocate: "I'm not any happier with the Turing Test as a test for thinking machines than I am with the Imitation Game [the Turing Test] as a test for femi-
ninity" (Hofstadter, 1985, p. 495). Certainly, even if a man could imitate a
woman, perfectly, he would still be a man. Imitations are just that: imitations.
It is important to define and program mechanisms that generate intelligent behavior. Artificial intelligence should not seek to merely solve problems, but
should rather seek to solve the problem of how to solve problems.

1.4   NEURAL NETWORKS

A human may be described as an intelligent problem-solving machine. Singh
(1966, p. 1) suggested that "the search for synthetic intelligence must begin
with an inquiry into the origin of natural intelligence, that is, into the working
of our own brain, its sole creator at present." The idea of constructing an artificial brain or *neural network* has been proposed many times (e.g., McCulloch
and Pitts, 1943; Rosenblatt, 1957, 1962; Samuel, 1959; Block, 1963; and others).

The brain is an immensely complex network of neurons, synapses, axons,
dendrites, and so forth (Figure 1-2), a "mammoth automatic telephone ex-

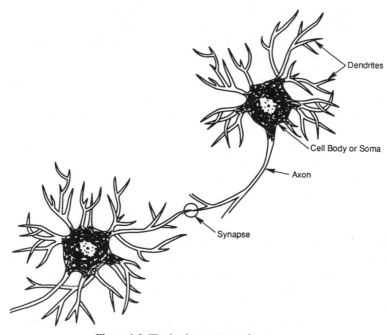

Figure 1-2 The basic structure of a neuron.

change" (Singh, 1966, p. 129). Through detailed modeling of these elements, a simulated network that is capable of diverse behaviors may be constructed. The human brain comprises at least 2×10^{10} neurons, each possessing about 10,000 synapses distributed over each dendritic tree with an average number of synapses on the axon of one neuron again being about 10,000 (Block, 1963; Palm, 1982, p. 10). Modeling the precise structure of this connection scheme would appear beyond the capabilities of foreseeable methods. Fortunately, this may not be necessary.

Rather than deduce specific replications of the human brain, models may be employed. Among the first such artificial neural network designs was the *perceptron* (Rosenblatt, 1957, 1958, 1960, 1962). A perceptron (Figure 1-3) consists of three types of units: sensory units, associator units, and response units. A stimulus will activate some sensory units. These sensory units in turn activate, with varying time delays and connection strengths, the associator units. These activations may be positive (excitatory) or negative (inhibitory). If the weighted sum of the activations at an associator unit exceeds a given threshold, the associator unit activates and sends a pulse, again weighted by a connection strength, onto the response units. There is obvious analogous behavior of units and neurons, of connections and axons and dendrites. The characteristics of the stimulus-response (input-output) of the perceptron describe its behavior.

Earlier work by Hebb (1949) indicated that neural networks could learn to recognize patterns by weakening and strengthening the connections between neurons. Rosenblatt (1957, 1960, 1962) and others (e.g., Keller, 1961; Kesler, 1961; Block, 1962; Block et al., 1962) studied the effects of changing the connection strengths in a perceptron by various rules (Rumelhart and McClelland, 1986, p. 155). Block (1962) indicated that when the perceptron was employed on some simple pattern recognition problems, the behavior of the machine degraded gradually with the removal of association units. That is, the perceptrons

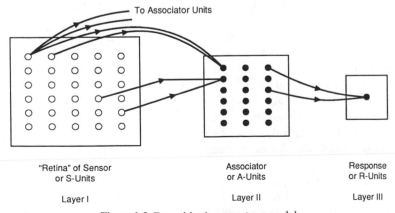

Figure 1-3 Rosenblatt's perceptron model.

were robust, not brittle. Rosenblatt (1962, p. 28) admitted that his perceptrons were "extreme simplifications of the central nervous system, in which some properties are exaggerated and others suppressed." But he also noted that the strength of the perceptron approach lay in the ability to analyze the model.

Minsky and Papert (1969) studied the computational limits of perceptrons with one layer of modifiable connections. They demonstrated that such processing units were not able to calculate mathematical functions such as parity or the topological function of connectedness without using an absurdly large number of predicates (Rumelhart and McClelland, 1986, p. 111). But these limitations do not apply to networks of perceptrons that consist of multiple layers of perceptrons, nor does their analysis address networks with recurrent feedback connections (rather than simple feedforward connections, although any perceptron with feedback can be approximated by an equivalent but larger feedforward network). Nevertheless, Minsky and Papert (1969, pp. 231–232) speculated that the study of multilayered perceptrons would be "sterile" in the absence of an algorithm to usefully adjust the connections of such architectures.

Minsky and Papert (1969, p. 4) offered, "We have agreed to use the name 'perceptron' in recognition of the pioneer work of Frank Rosenblatt," but Block (1970) noted that "they study a severely limited class of machines from a viewpoint quite alien to Rosenblatt's." While Block (1970) recognized the mathematical prowess of Minsky and Papert, he also replied, "Work on the four-layer *Perceptrons* has been difficult, but the results suggest that such systems may be rich in behavioral possibilities." Block (1970) admitted the inability of simple perceptrons to perform functions such as parity checking and connectedness, but remarked, "Human beings cannot perceive the parity of large sets . . . nor connectedness." The recognition of more common objects such as faces was viewed as a more appropriate test.

Block questioned prophetically, "Will the formulations or methods developed in the book have a serious influence on future research in pattern recognition, threshold logic, psychology, or biology; or will this book prove to be only a monument to the mathematical virtuosity of Minsky and Papert? We shall have to wait for a few years to find out" (Block, 1970).

Minsky and Papert's speculation that efforts with multilayered perceptrons would be sterile served in part to restrict funding and thus research efforts in neural networks during the 1970s and early 1980s. The criticisms by Minsky and Papert (1969) were passionate and persuasive. Papert (1988) admitted, "Yes, there was *some* hostility in the energy behind the research reported in *Perceptrons,* and there is *some* degree of annoyance at the way the new [resurgence in neural network research] has developed; part of our drive came, as we quite plainly acknowledged in our book, from the fact that funding and research energy were being dissipated on what still appcar to me . . . to be misleading attempts to use connectionist methods in practical applications." Subsequent to Minsky and Papert (1969), neural network research was continued by Grossberg (1976, 1982), Amari (1967, 1971, 1972, and many others), Kohonen (1984), and others, but to a lesser degree than that conducted on knowledge-based systems. A resurgence of interest grew in the mid-1980s following further research by Hopfield (1982), Hopfield and Tank (1985), Rumelhart and McClelland (1986), Mead (1989), and others (Simpson, 1990, pp. 136–145, provides a concise review of efforts in neural networks; also, see Hecht-Nielsen, 1990; Haykin, 1994).

It is now well known that multiple layers of perceptrons with variable connection strengths, bias terms, and nonlinear sigmoid functions can approximate arbitrary measurable mapping functions. In fact, universal function approximators can be constructed with a single hidden layer of squashing units and an output layer of linear units (Cybenko, 1989; Hornik et al., 1989; Barron, 1993). The application of such structures to pattern recognition problems is now routine. But as noted, such structures are simply mapping functions. Functions are not intrinsically intelligent. The crucial problem then becomes training such functions to yield the appropriate stimulus-response. There are several proposed methods to discover a suitable set of weights and bias terms, given an overall topological structure and a set of previously classified patterns (e.g., Werbos, 1974; Hopfield, 1982; Rumelhart and McClelland, 1986; Arbib and Hanson, 1990; Levine, 1991).

Currently, the method most often employed (i.e., back propagation) is based on a simple gradient descent search of the error response surface determined by the set of weights and biases. But this method is likely to discover suboptimal solutions because the response surface is a general nonlinear function and may possess many local optima. Moreover, a gradient descent algorithm is inherently no more intelligent than any other deterministic algorithm. Conceptually, the back propagation routine is little different from the addition of two integers.

There does not appear to be any evidence to suggest that biologic neural networks alter connection strengths via methods similar to gradient-based search. If the bottom-up approach to neural networks is to lead to artificially intelligent machines, the intelligence must come in the search for the appropriate structure and parameters of the particular mapping function. "Learning models which cannot adaptively cope with unpredictable changes in a complex environment have an unpromising future as models of mind and brain" (Grossberg, 1987). Models of self-organizing networks (e.g., Grossberg, 1987, and others) appear more broadly useful as explanations of biologic nervous systems.

1.5   DEFINITION OF INTELLIGENCE

If one word were to be used to describe research in artificial intelligence, that word might be *fragmented*. Opinions as to the cause of this scattered effort are varied. Atmar (1976) remarked:

> Perhaps the major problem is our viewpoint. Intelligence is generally regarded as a uniquely human quality. And yet we, as humans, do not understand ourselves, our capabilities, or our origins of thought. In our rush to catalogue and emulate our own staggering array of behavioral responses, it is only logical to suspect that investigations into the primal causative factors of intelligence have been passed over in order to more rapidly obtain the immediate consequences of intelligence.

Minsky (1991), on the other hand, assigns blame to attempts at unifying theories of intelligence: "There is no one best way to represent knowledge or to solve problems, and the limitations of current machine intelligence largely stem from seeking unified theories or trying to repair the deficiencies of theoretically neat but conceptually impoverished ideological positions."

A prerequisite to embarking upon research in artificial intelligence should be a definition of the term *intelligence*. As noted above (Atmar, 1976), many definitions of intelligence have relied on this property being uniquely human (e.g., Singh, 1966, p. 1) and often reflect a highly anthropocentric view. Schank (1984, p. 49) stated:

> No question in AI has been the subject of more intense debate than that of assessing machine intelligence and understanding. People unconsciously feel that calling something other than humans intelligent denigrates humans and reduces their vision of themselves as the center of the universe. Dolphins are intelligent. Whales are intelligent. Apes are intelligent. Even dogs and cats are intelligent.

Certainly other living systems can be described as being intelligent, without ascribing specific intelligence to any individual member of the system. Any proposed definition of intelligence should not rely on comparisons to individual organisms.

In contrast to Staugaard's comments (1987, p. 23), Minsky has offered the following definition of intelligence (1985, p. 71): "Intelligence . . . means . . . the ability to solve hard problems." But how hard does a problem have to be? Who is to decide which problem is hard? All problems are hard until you know how to solve them, at which point they become easy. Finding the slope of a polynomial at any specific point is very difficult, unless you are familiar with derivatives, in which case it is trivial. Such a definition would appear problematic.

For an organism, or any system, to be intelligent, it must make decisions. Any decision may be described as the selection of how to allocate the available resources. And an intelligent system must face a range of decisions, for if there were only one possible decision, there would really be no decision at all. Moreover, decision making requires a goal. Without the existence of a goal, decision making is pointless. The intelligence of such a decision-making entity becomes a meaningless quality.

This argument begs the question, "Where do goals come from?" Consider biologically reproducing organisms. They exist in a finite arena; as a consequence, there is competition for the available resources. Natural selection is inevitable in any system of self-replicating organisms that fill the available resource space. Selection stochastically eliminates those variants that do not acquire sufficient resources. Thus, while evolution as a process is purposeless, the first purposeful goal imbued into all living systems is survival. Those variants that do not exhibit behaviors that meet this goal are stochastically culled. The genetically preprogrammed behaviors of the survivors (and thus the goal of survival) are reinforced in every generation through intense competition.

Such a notion has been suggested many times. For example, Carne (1965, p. 3) remarked, "Perhaps the basic attribute of an intelligent organism is its capability to learn to perform various functions within a changing environment so as to survive and to prosper." Atmar (1976) offered, "Intelligence is that property in which the organism senses, reacts to, learns from, and subsequently adapts its behavior to its present environment in order to better promote its own survival."

Note that any automaton whose behavior (i.e., stimulus-response pairs that depend on the state of the organism) is completely prewired (e.g., a simple hand-held calculator or the hunting wasp described previously) cannot learn anything. Nor can it make decisions. Such systems should not be viewed as intelligent. But this should not be taken as a contradiction to the statement that "the genetically preprogrammed behaviors of the survivors (and thus the goal of survival) are passed along to future progeny." Behaviors in all biota, individuals or populations, are dependent on underlying genetic programs. In some cases, these programs mandate specific behaviors; in others, they create nervous systems capable of adapting behavior of the organism based on its experiences.

But the definition of intelligence should not be restricted to biological organisms. Intelligence is a property of purpose-driven decision makers. It applies equally well to humans, colonies of ants, robots, social groups, and so forth. Thus, more generally, following Fogel (1964; Fogel et al., 1966, p. 2), intelligence may be defined as *the capability of a system to adapt its behavior to meet its goals in a range of environments*. For species, survival is a necessary goal in any given environment; for a machine, both goals and environments may be imbued by the machine's creators.

1.6 INTELLIGENCE, THE SCIENTIFIC METHOD, AND EVOLUTION

Ornstein (1965) argued that all learning processes are adaptive. The most important aspect of such learning processes is the "development of implicit or explicit techniques to accurately estimate the probabilities of future events." Similar notions have been offered by Atmar (1976, 1979). When faced with a changing environment, the adaptation of behavior becomes little more than a shot in the dark if the system is incapable of predicting future events. Ornstein (1965) suggested that as predicting future events is the "forte of science," it is sensible to examine the scientific method for useful cues in the search for effective learning techniques.

The scientific method (Figure 1-4) is an iterative process that facilitates the gaining of new knowledge about the underlying processes of an observable environment. Unknown aspects of the environment are estimated. Data are collected in the form of previous observations or known results and combined with newly acquired measurements. After the removal of known erroneous data, a class of models of the environment that is consistent with the data is generalized. This process is necessarily inductive. The class of models is then generally reduced by parametrization, a deductive process. The specific hypothesized model (or models) is then tested in its ability to predict future aspects of the environment. Models that prove worthy are modified, extended, or combined to form new hypotheses that carry on a "heredity of reasonableness" (Fogel, 1964). This process is iterated until a sufficient level of credibility is achieved. "As the hypotheses correspond more and more closely with the logic of the environment they provide an 'understanding' that is demonstrated in terms of improved goal-seeking behavior in the face of that environment" (Fogel et al., 1966, p. 111). It appears reasonable to seek methods to mechanize the scientific method in an algorithmic formulation so that a machine may carry out the procedure and similarly gain knowledge about its environment and adapt its behavior to meet goals.

The scientific method can be used to describe a process of human investigation of the universe, or of learning processes in general. Atmar (1976), following Weiner (1961), proposed that there are "three distinct organization

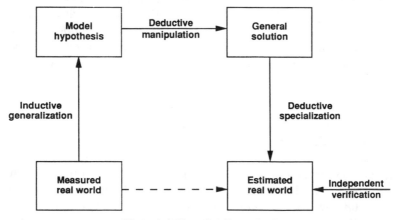

Figure 1-4 The scientific method.

forms of intelligence: phylogenetic, ontogenetic and sociogenetic, which are equivalent to one another in process, each containing a quantum unit of mutability and a reservoir of learned behavior." Individuals of most species are capable of learning ontogenetically (self-arising within the individual). The minimum unit of mutability is the proclivity of a neuron to fire. The reservoir of learned behavior becomes the entire "collection of engrams reflecting the sum of knowledge the organism possesses about its environment" (Atmar, 1976). Sociogenetic learning (arising within the group) is the basis for a society to acquire knowledge and communicate (Wilson, 1971; Atmar, 1976). The quantum unit of mutability is the "idea," while "culture" is the reservoir of learned behavior.

But phylogenetic learning (arising from within the lineage) is certainly the most ancient, and the most commonly exhibited, form of intelligence. The quantum unit of mutability is the nucleotide base pair, and the reservoir of learned behavior is the genome of the species. The recognition of evolution as an intelligent learning process is a recurring idea (Cannon, 1932; Turing, 1950; Fogel, 1962; and others). Fogel et al. (1966, p. 112) developed a correspondence between natural evolution and the scientific method. In nature, individual organisms serve as hypotheses concerning the logical properties of their environment. Their behavior is an inductive inference concerning some as yet unknown aspects of that environment. Validity is demonstrated by their survival. Over successive generations, organisms become successively better predictors of their surroundings.

Minsky (1985, p. 71) disagreed, claiming that evolution is not intelligent "because people also use the word 'intelligence' to emphasize swiftness and efficiency. Evolution's time rate is so slow that we don't see it as intelligent, even though it finally produces wonderful things we ourselves cannot yet make." But time does not enter into the definition of intelligence offered above, and it need not. Atmar (1976) admits that "the learning period [for evolution] may be tortuously long by human standards, but it is real, finite, and continuous." And evolution can proceed quite rapidly (e.g., viruses). Units of time are a human creation. The simulation of the evolutionary process on a computer need not take billions of years. Successive generations can be interated very quickly. Metaphorically, the videotape of evolution can be played at fast forward with no alteration of the basic algorithm. Arguments attacking the speed of the process (as opposed to the rate of learning) are without merit.

There is obvious value in plasticity. Organisms that can adapt to changes in the environment at a greater rate than through direct physical modifications will tend to outcompete less mutable organisms. Thus the evolutionary benefit of ontogenetic learning is obvious. Sociogenetic learning is even more powerful as the communicative group possesses even greater plasticity in behavior, a more durable memory, and a greater range of possible mutability (Atmar, 1976). But both ontogenetic learning and sociogenetic learning are "tricks" of phylogenetic learning, invented through the randomly driven search of alternative methods of minimizing behavioral surprise to the evolving species.

If the evolutionary process is accepted as being fundamentally analogous to the scientific method, then so must the belief that this process can be mechanized and programmed on a computing machine. Evolution, like all other natural processes, is a mechanical procedure operating on and within the laws of physics and chemistry (Fogel, 1964; Fogel et al., 1966; Wooldridge, 1968, p. 25; Atmar, 1976, 1979, 1991; Mayr, 1988, pp. 148–159). If the scientific method is captured in an algorithm, then so must induction be captured as an intrinsic part of that algorithm.

Fogel et al. (1966, p. 122) noted that induction had been presumed to require creativity and imagination, but through the simulation of evolution, induction can be reduced to a routine procedure. Such notions have a propensity to generate pointed responses. Lindsay (1968) wrote in criticism of Fogel et al. (1966), "The penultimate indignity is a chapter in which the scientific method is described as an evolutionary process and hence mechanizable with the procedures described." People unconsciously feel that calling something not human "intelligent" denigrates humans (Schank, 1984, p. 49; cf. Pelletier, 1978, pp. 240–241). Yet wishing something away does not make it so. Wooldridge (1968, p. 129) stated:

> In particular, it must not be imagined that reduction of the processes of intelligence to small-step mechanical operations is incompatible with the apparently spontaneous appearance of new and original ideas to which we apply such terms as "inspiration," "insight," or "creativity." To be sure, there is no way for the physical methods . . . to produce full-blown thoughts or ideas from out of the blue. But it will be recalled that there is a solution for this problem. The solution is to deny that such spontaneity really exists. The argument is that this is an example of our being led astray by attributing too much reality to our subjective feelings—that the explanation of the apparent freedom of thought is the incompleteness of our consciousness and our resulting lack of awareness of the tortuous . . . nature of our thought processes.

Hofstadter (1985, p. 529) went further:

> Having creativity is an automatic consequence of having the proper representation of concepts in a mind. It is not something you add on afterward. It is built into the way concepts are. To spell this out more concretely: If you have succeeded in making an accurate model of *concepts*, you have thereby also succeeded in making a model of the creative process, and even of consciousness.

Creativity and imagination are part of the invention of evolution just as are eyes, opposable thumbs, telephones, and calculators.

The process of evolution can be described as four essential processes: self-reproduction, mutation, competition, and selection (Mayr, 1988; Hoffman,

1989; and many others). The self-reproduction of germline DNA and RNA systems is well known. In a positively entropic universe (as dictated by the second law of thermodynamics), the property of mutability is guaranteed; error in information transcription is inevitable. A finite arena guarantees the existence of competition. Selection becomes the natural consequence of the excess of organisms that have filled the available resource space (Atmar, 1979). The implication of these very simple rules is that evolution is a procedure that can be simulated and used to generate creativity and imagination mechanically.

1.7 EVOLVING ARTIFICIAL INTELLIGENCE

It is natural to conclude that by simulating the evolutionary learning process on a computer, the machine can become intelligent, that it can adapt its behavior to meet goals in a range of environments. "Intelligence is a basic property of life" (Atmar, 1976). It has occurred at the earliest instance of natural selection and has pervaded all subsequent living organisms. In many ways, life is intelligence, and the processes cannot be easily separated.

Numerous opinions about the proper goal for artificial intelligence research have been expressed. But intuitively, intelligence must be the same process in living organisms as it is in machines. The "artificial" is not nearly as important as the "intelligence." "Artificial Intelligence is the study of intelligent behavior. Its ultimate goal is a theory of intelligence that accounts for the behavior of naturally occurring intelligent entities and that guides the creation of artificial entities capable of intelligent behavior" (Genesereth and Nilsson, 1987). Evolutionary processes account for such intelligent behavior and can be simulated and used for the creation of intelligent machines.

REFERENCES

Adams, J. B. (1976). "A Probability Model of Medical Reasoning and the MYCIN Model," *Mathematical Biosciences,* Vol. 32, pp. 177–186.

Amari, S.-I. (1967). "A Theory of Adaptive Pattern Classifiers," *IEEE Trans. of Elec. Comp.,* Vol. EC-16, pp. 299–307.

Amari, S.-I. (1971). "Characteristics of Randomly Connected Threshold-Element Networks and Network Systems," *Proc. of the IEEE,* Vol. 59:1, pp. 35–47.

Amari, S.-I. (1972). "Learning Patterns and Pattern Sequences by Self-Organizing Nets of Threshold Elements," *IEEE Trans. Comp.,* Vol. C-21, pp. 1197–1206.

Arbib, M. A., and A. R. Hanson (1990). *Vision, Brain, and Cooperative Computation.* Cambridge, MA: MIT Press.

Atmar, J. W. (1976). "Speculation on the Evolution of Intelligence and Its Possible Realization in Machine Form." Sc.D. diss., New Mexico State University, Las Cruces.

Atmar, J. W. (1979). "The Inevitability of Evolutionary Invention." Unpublished manuscript.

Atmar, W. (1991). "On the Role of Males," *Animal Behaviour,* Vol. 41, pp. 195–205.

Barr, A., and E. A. Feigenbaum (1981). *The Handbook of Artificial Intelligence,* Vol. 1. San Mateo, CA: William Kaufmann.

Barr, A., P. R. Cohen, and E. A. Feigenbaum (1989). *The Handbook of Artificial Intelligence,* Vol. 4. Reading, MA: Addison-Wesley.

Barron, A. R. (1993). "Universal Approximation Bounds for Superpositions of a Sigmoidal Function," *IEEE Trans. Info. Theory,* Vol. 39:3, pp. 930–945.

Bennett, J. S., and C. R. Hollander (1981). "DART: An Expert System for Computer Fault Diagnosis," *Proc. of IJCAI-81,* pp. 843–845.

Bernstein, A., V. De, M. Roberts, T. Arbuckle, and M. A. Belsky (1958). "A Chess Playing Program for the IBM 704," *Proc. West Jut. Comp. Conf.,* Vol. 13, pp. 157–159.

Bezdek, J. C., and S. K. Pal (1992). *Fuzzy Models for Pattern Recognition: Models that Search for Structures in Data.* Piscataway, NJ: IEEE Press.

Block, H. D. (1962). "The Perceptron: A Model for Brain Functioning," *Rev. Mod. Phys.,* Vol. 34, pp. 123–125.

Block, H. D. (1963). "Adaptive Neural Networks as Brain Models," *Proc. of Symp. Applied Mathematics,* Vol. 15, pp. 59–72.

Block, H. D. (1970). "A Review of 'Perceptrons: An Introduction to Computational Geometry,'" *Information and Control,* Vol. 17:5, pp. 501–522.

Block, H. D., B. W. Knight, and F. Rosenblatt (1962). "Analysis of a Four Layer Series Coupled Perceptron," *Rev. Mod. Phys.,* Vol. 34, pp. 135–142.

Buchanan, B. G., and E. H. Shortliffe (1985). *Rule-Based Expert Systems: The MYCIN Experiments of the Stanford Heuristic Programming Project.* Reading, MA: Addison-Wesley.

Campbell, A. N., V. F. Hollister, R. O. Duda, and P. E. Hart (1982). "Recognition of a Hidden Mineral Deposit by an Artificial Intelligence Program," *Science,* Vol. 217, pp. 927–929.

Cannon, W. D. (1932). *The Wisdom of the Body.* New York: Norton and Company.

Carne, E. B. (1965). *Artificial Intelligence Techniques.* Washington, DC: Spartan Books.

Cerf, C., and V. Navasky (1984). *The Experts Speak: The Definitive Compendium of Authoritative Misinformation.* New York: Pantheon Books.

Charniak, E., and D. V. McDermott (1985). *Introduction to Artificial Intelligence.* Reading, MA: Addison-Wesley.

Clark, D. (1997). "Deep Thoughts on Deep Blue," *IEEE Expert,* Vol. 12:4, p. 31.

Countess of Lovelace (1842). "Translator's Notes to an Article on Babbage's Analytical Engine." In *Scientific Memoirs,* Vol. 3, edited by R. Taylor, pp. 691–731.

Covington, M. A., D. Nute, and A. Vellino (1988). *Prolog Programming in Depth.* Glenview, IL: Scott, Foresman.

Cybenko, G. (1989). "Approximations by Superpositions of a Sigmoidal Function," *Math. Contr. Signals, Syst.,* Vol. 2, pp. 303–314.

Dreyfus, H., and S. Dreyfus (1984). "Mindless Machines: Computers Don't Think Like Experts, and Never Will," *The Sciences,* November/December, pp. 18–22.

Dreyfus, H., and S. Dreyfus (1986). "Why Computers May Never Think Like People," *Tech. Review,* January, pp. 42–61.

Dubois, D., and H. Prade (1982). "A Class of Fuzzy Measures Based on Triangular Norms—A General Framework for the Combination of Uncertain Information," *Int. J. of General Systems,* Vol. 8:1.

Duda, R., P. E. Hart, N. J. Nilsson, P. Barrett, J. G. Gaschnig, K. Konolige, R. Reboh, and J. Slocum (1978). "Development of the PROSPECTOR Consultation System for Mineral Exploration," SRI Report, Stanford Research Institute, Menlo Park, CA.

Feigenbaum, E. A., B. G. Buchanan, and J. Lederberg (1971). "On Generality and Problem Solving: A Case Study Involving the DENDRAL Program." In *Machine Intelligence 6,* edited by B. Meltzer and D. Michie. New York: American Elsevier, pp. 165–190.

Fogel, L. J. (1962). "Autonomous Automata," *Industrial Research,* Vol. 4, pp. 14–19.

Fogel, L. J. (1964). "On the Organization of Intellect." Ph.D. diss., UCLA.

Fogel, L. J., A. J. Owens, and M. J. Walsh (1966). *Artificial Intelligence through Simulated Evolution.* New York: John Wiley.

Genesereth, M. R., and N. J. Nilsson (1987). *Logical Foundations of Artificial Intelligence.* Los Altos, CA: Morgan Kaufmann.

Gould, J. L., and C. G. Gould (1985). "An Animal's Education: How Comes the Mind to be Furnished?" *The Sciences,* Vol. 25:4, pp. 24–31.

Greenblatt, R., D. Eastlake, and S. Crocker (1967). "The Greenblatt Chess Program," *FJCC,* Vol. 31, pp. 801–810.

Grossberg, S. (1976). "Adaptive Pattern Classification and Universal Recoding: Part I. Parallel Development and Coding of Neural Feature Detectors," *Biological Cybernetics,* Vol. 23, pp. 121–134.

Grossberg, S. (1982). *Studies of Mind and Brain.* Dordrecht, Holland: Reidel.

Grossberg, S. (1987). "Competitive Learning: From Interactive Activation to Adaptive Resonance," *Cognitive Science,* Vol. 11, pp. 23–63.

Haykin, S. (1994). *Neural Networks: A Comprehensive Foundation.* New York: Macmillan.

Hebb, D. O. (1949). *The Organization of Behavior.* New York: John Wiley.

Hecht-Nielsen, R. (1990). *Neurocomputing.* Reading, MA: Addison-Wesley.

Hoffman, A. (1989). *Arguments on Evolution: A Paleontologist's Perspective.* New York: Oxford Univ. Press.

Hofstadter, D. R. (1985). *Metamagical Themas: Questing for the Essence of Mind and Pattern.* New York: Basic Books.

Hopfield, J. J. (1982). "Neural Networks and Physical Systems with Emergent Collective Computational Abilities," *Proc. Nat. Acad. of Sciences,* Vol. 79, pp. 2554–2558.

Hopfield, J. J., and D. Tank (1985). " 'Neural' Computation of Decision in Optimization Problems," *Biological Cybernetics,* Vol. 52, pp. 141–152.

Hornik, K., M. Stinchcombe, and H. White (1989). "Multilayer Feedforward Networks Are Universal Approximators," *Neural Networks,* Vol. 2, pp. 359–366.

Jackson, P. C. (1985). *Introduction to Artificial Intelligence,* 2nd ed. New York: Dover.

Kandel, A. (1986). *Fuzzy Expert Systems.* Boca Raton, FL: CRC Press.

Keller, H. B. (1961). "Finite Automata, Pattern Recognition and Perceptrons," *J. Assoc. Comput. Mach.,* Vol. 8, pp. 1–20.

Kesler, C. (1961). "Preliminary Experiments on Perceptron Applications to Bubble Chamber Event Recognition." Cognitive Systems Research Program, Rep. No. 1, Cornell University, Ithaca, NY.

Kohonen, T. (1984). *Self-Organization and Associative Memory.* Berlin: Springer-Verlag.

Lenat, D. B. (1983). "EURISKO: A Program that Learns New Heuristics and Domain Concepts," *Artificial Intelligence,* Vol. 21, pp. 61–98.

Levine, D. S. (1991). *Introduction to Neural and Cognitive Modeling.* Hillsdale, NJ: Lawrence Erlbaum.

Levy, D. N. L., and M. Newborn (1991). *How Computers Play Chess.* New York: Computer Science Press.

Lindsay, R. K. (1968). "Artificial Evolution of Intelligence," *Contemp. Psych.,* Vol. 13:3, pp. 113–116.

Lindsay, R. K., B. G. Buchanan, E. A. Feigenbaum, and J. Lederberg (1980). *Applications of Artificial Intelligence for Organic Chemistry: The DENDRAL Project.* New York: McGraw-Hill.

Mayr, E. (1988). *Toward a New Philosophy of Biology: Observations of an Evolutionist.* Cambridge, MA: Belknap Press.

McCarthy, J., P. J. Abrahams, D. J. Edwards, P. T. Hart, and M. I. Levin (1962). *LISP 1.5 Programmer's Manual.* Cambridge, MA: MIT Press.

McCarthy, J. (1997). "AI as Sport," *Science,* Vol. 276, pp. 1518–1519.

McCulloch, W. S., and W. Pitts (1943). "A Logical Calculus of the Ideas Immanent in Nervous Activity," *Bull. Math. Biophysics,* Vol. 5, pp. 115–133.

Mead, C. (1989). *Analog VLSI and Neural Systems.* Reading, MA: Addison-Wesley.

Minsky, M. L. (1985). *The Society of Mind.* New York: Simon and Schuster.

Minsky, M. L. (1991). "Logical Versus Analogical or Symbolic versus Connectionist or Neat versus Scruffy," *AI Magazine,* Vol. 12:2, pp. 35–51.

Minsky, M. L., and S. Papert (1969). *Perceptrons.* Cambridge, MA: MIT Press.

Negoita, C. V., and D. Ralescu (1987). *Simulation, Knowledge-Based Computing, and Fuzzy Statistics.* New York: Van Nostrand Reinhold.

Newell, A., J. C. Shaw, and H. A. Simon (1958). "Chess Playing Programs and the Problem of Complexity," *IBM J. of Res. & Dev.,* Vol. 4:2, pp. 320–325.

Newell, A., and H. A. Simon (1963). "GPS: A Program that Simulates Human Thought." In *Computers and Thought,* edited by E. A. Feigenbaum and J. Feldman. New York: McGraw-Hill, pp. 279–293.

Ornstein, L. (1965). "Computer Learning and the Scientific Method: A Proposed

Solution to the Information Theoretical Problem of Meaning," *J. of the Mt. Sinai Hosp.*, Vol. 32:4, pp. 437–494.

Palm, G. (1982). *Neural Assemblies: An Alternative Approach to Artificial Intelligence.* Berlin: Springer-Verlag.

Papert, S. (1988). "One AI or Many?" In *The Artificial Intelligence Debate: False Starts, Real Foundations,* edited by S. R. Braubard. Cambridge, MA: MIT Press.

Pelletier, K. R. (1978). *Toward a Science of Consciousness.* New York: Dell.

Rich, E. (1983). *Artificial Intelligence.* New York: McGraw-Hill.

Rosenblatt, F. (1957). "The Perceptron, a Perceiving and Recognizing Automaton." Project PARA, Cornell Aeronautical Lab. Rep., No. 85-640–1, Buffalo, NY.

Rosenblatt, F. (1958). "The Perceptron: A Probabilistic Model for Information Storage and Organization in the Brain," *Psychol. Rev.,* Vol. 65, p. 386.

Rosenblatt, F. (1960). "Perceptron Simulation Experiments," *Proc. IRE,* Vol. 48, pp. 301–309.

Rosenblatt, F. (1962). *Principles of Neurodynamics: Perceptrons and the Theory of Brain Mechanisms.* Washington, DC: Spartan Books.

Rumelhart, D. E. and J. L. McClelland (1986). *Parallel Distributed Processing: Explorations in the Microstructures of Cognition,* Vol. 1. Cambridge, MA: MIT Press.

Samuel, A. L. (1959). "Some Studies in Machine Learning Using the Game of Checkers," *IBM J. of Res. and Dev.,* Vol. 3:3, pp. 210–229.

Samuel, A. L. (1963). "Some Studies in Machine Learning Using the Game of Checkers." In *Computers and Thought,* edited by E. A. Feigenbaum and J. Feldman. New York: McGraw-Hill, pp. 71–105.

Schaeffer, J. (1996). *One Jump Ahead: Challenging Human Supremacy in Checkers.* Berlin: Springer.

Schank, R. C. (1984). *The Cognitive Computer: On Language, Learning, and Artificial Intelligence.* Reading. MA: Addison-Wesley.

Schildt, H. (1987). *Artificial Intelligence Using C.* New York: McGraw-Hill.

Shannon, C. E. (1950). "Programming a Computer for Playing Chess," *Philosophical Magazine,* Vol. 41, pp. 256–275.

Shortliffe, E. H. (1974). "MYCIN: A Rule-Based Computer Program for Advising Physicians Regarding Antimicrobial Therapy Selection." Ph.D. diss., Stanford University.

Simpson, P. K. (1990). *Artificial Neural Systems: Foundations, Paradigms, Applications and Implementations.* New York: Pergamon Press.

Singh, J. (1966). *Great Ideas in Information Theory, Language and Cybernetics.* New York: Dover.

Staugaard, A. C. (1987). *Robotics and AI: An Introduction to Applied Machine Intelligence.* Englewood Cliffs, NJ: Prentice Hall.

Swade, D. D. (1993). "Redeeming Charles Babbage's Mechanical Computer," *Scientific American,* Vol. 268, No. 2, February, pp. 86–91.

Turing, A. M. (1950). "Computing Machinery and Intelligence," *Mind,* Vol. 59, pp. 433–460.

Turing, A. M. (1953). "Digital Computers Applied to Games." In *Faster than Thought,* edited by B. V. Bowden. London: Pittman, pp. 286–310.

Waterman, D. A. (1986). *A Guide to Expert Systems.* Reading, MA: Addison-Wesley.

Weiner, N. (1961). *Cybernetics,* Part 2. Cambridge, MA: MIT Press.

Werbos, P. (1974). "Beyond Regression: New Tools for Prediction and Analysis in the Behavioral Sciences." Ph.D. diss., Harvard University.

Wilson, E. O. (1971). *The Insect Societies.* Cambridge, MA: Belknap Press.

Wooldridge, D. E. (1968). *The Mechanical Man: The Physical Basis of Intelligent Life.* New York: McGraw-Hill.

Yager, R. R. (1980). "A Measurement-Informational Discussion of Fuzzy Union and Intersection," *IEEE Trans. Syst. Man Cyber.,* Vol. 10:1, pp. 51–53.

Zadeh, L. (1965). "Fuzzy Sets," *Information and Control,* Vol. 8, pp. 338–353.

CHAPTER 2

NATURAL EVOLUTION

2.1 THE NEO-DARWINIAN PARADIGM

A revolution in biological thought, and indeed in human philosophy, was begun when Charles Darwin and Alfred Russel Wallace each presented their evidence for the theory of evolution before the Linnean Society of London on July 1, 1858. The position that their theories and observations have come to occupy in biology cannot be exaggerated. Dobzhansky has remarked, "Nothing in biology makes sense except in the light of evolution" (Dobzhansky et al., 1977, frontispiece). And Mayr (1963, p. 1) has said, "There is no area of biology in which [evolution] has not served as an ordering principle." Classic Darwinian evolutionary theory, combined with the selectionism of Weismann and the genetics of Mendel, has now become a rather universally accepted set of arguments known as the neo-Darwinian paradigm (Fisher, 1930; Wright, 1931; Haldane, 1932; Dobzhansky, 1937; Huxley, 1942; Mayr, 1942; Simpson, 1944; Rensch, 1947; Stebbins, 1950).

Neo-Darwinism asserts that the history of the vast majority of life is fully accounted for by only a very few statistical processes acting on and within populations and species (Hoffman, 1989, p. 39). These processes are reproduction, mutation, competition, and selection. Reproduction is an obvious property of all life. But similarly as obvious, mutation is guaranteed in any system that continuously reproduces itself in a positively entropic universe. Competition and selection become the inescapable consequences of any expanding population constrained to a finite arena. Evolution is then the result of these fundamental interacting stochastic processes as they act on populations, generation after generation (Huxley, 1963; Wooldridge, 1968, p. 25; and others).

The questions that arise if evolution is to be accurately emulated for engineering purposes are many. (1) What is truly being optimized by evolution? (2) How extensive is that optimization? (3) To whom do the benefits of optimization accrue: to the individual gene, to the individual (and its genome), or to the species as a whole? (4) What is the true nature of the gene? (5) What is the significance of neutral mutations in evolution? (6) What are the benefits of sex and sexual selection? Evolutionary mechanisms create a complex web of interacting effects that cannot be easily partitioned or disentangled (Hoffman, 1989, pp. 32–33), but the physics that governs the evolutionary process may be quite simple and expressible by a rather small set of rules.

2.2 THE GENOTYPE AND THE PHENOTYPE: THE OPTIMIZATION OF BEHAVIOR

Living organisms can be viewed as a duality of their genotype (the underlying genetic coding) and their phenotype (the manner of response contained in the behavior, physiology and morphology of the organism). Lewontin (1974) illustrated this distinction by specifying two state spaces: a populational genotypic (informational) space $\mathbf{G}$ and a populational phenotypic (behavioral) space $\mathbf{P}$. Four functions map elements in $\mathbf{G}$ and $\mathbf{P}$ to each other (Figure 2-1). Atmar (1992) modified these functions to be

$$f_1: \mathbf{I} \times \mathbf{G} \to \mathbf{P},$$

$$f_2: \mathbf{P} \to \mathbf{P},$$

$$f_3: \mathbf{P} \to \mathbf{G},$$

$$f_4: \mathbf{G} \to \mathbf{G}.$$

The function f_1, *epigenesis*, maps the element $g_1 \in \mathbf{G}$ into the phenotypic space $\mathbf{P}$ as a particular collection of phenotypes p_1 whose development is modified by its environment, an indexed set of symbols $(i_1, \ldots, i_k) \in \mathbf{I}$, where $\mathbf{I}$ is the set of all such environmental sequences. The function f_2, *selection*, maps phenotypes p_1 into p_2. As natural selection only operates on the phenotypic expressions of the genotype (Mayr, 1960, p. 24, 1970, p. 131; Hartl and Clark, 1989, p. 431), the underlying coding g_1 is not involved in function f_2. The function f_3, *genotypic survival*, describes the effects of selection and migration processes on $\mathbf{G}$. Function f_4, *mutation*, maps the representative codings $g_2 \in \mathbf{G}$ to the point $g_1' \in \mathbf{G}$. This function represents the "rules" of mutation and recombination, and encompasses all genetic changes. With the creation of the new population of genotypes g_1', one generation is complete. Evolutionary adaptation occurs over successive iterations of these mappings.

The neo-Darwinian argument asserts that natural selection is the predominant mediating evolutionary force that prevails in shaping the phenotypic characters of organisms in the vast majority of situations encountered in nature (Mayr, 1988; Hoffman, 1989; Brunk, 1991; Wilson, 1992, p. 75; and others). It is strictly an a posteriori process that rewards current success (Mayr, 1988, p. 43) primarily through the statistical culling of inappropriate individu-

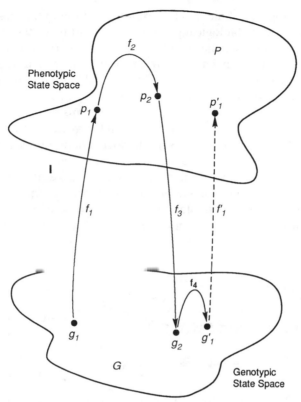

Figure 2-1 The evolution of a population within a single generation. Evolution can be viewed as occurring as a succession of four mapping functions (*epigenesis, selection, genotypic survival,* and *mutation*) relating the genotypic information state space and the phenotypic behavioral state space.

als. Selection acts in the face of phenotypic variation. Parents and their off-spring typically demonstrate at least a general resemblance in their phenotypic traits (Hartl and Clark, 1989, p. 430). Their behaviors may in fact be virtually identical, that is, reproduction may be very nearly replication (Hoffman, 1989, p. 24). But phenotypic variation is always observed within populations and species and is largely the result of mutation and recombination (if applicable), environmental constraints placed on an individual's development, and replicative errors of the genome. The interaction between the species and its environment (which includes other organisms) determines the relative success (differential reproduction) or failure (genetic death) of the species.

Selection is often viewed as leading to the maintenance or increase of populations' "fitness," where fitness is defined as the ability to survive and reproduce in a specific environment (Hartl and Clark, 1989, p. 147). It has been asserted that although fitness cannot be directly measured, "its distribution in a population can be roughly estimated in any given environmental context on the basis of ecology and functional morphology of the organisms. Hence it is an empirically testable biological proposition" (Hoffman, 1989, p. 29). Atmar (1994) indicated that a singular measure of evolutionary fitness is the appropriateness of the species' behavior in terms of its ability to anticipate its environment. The quantitative ability to perform suitable prediction and elicit appropriate response yields a measure of fitness.

In contrast, Fisher (1930) defined fitness in terms of the gene as the per capita rate of increase (of a genotype, based on the average effect of gene substitutions, e.g., Fisher, 1930, pp. 30–34). But such a definition is problematic in light of the pleiotropic and polygenic nature of the genotype. *Pleiotropy* is the effect that a single gene may simultaneously affect several phenotypic traits. *Polygeny* is the effect that a single phenotypic characteristic of an individual may be determined by the simultaneous interaction of many genes (Figure 2-2). Assigning fitness to individual genes implicitly ignores their interactions, or

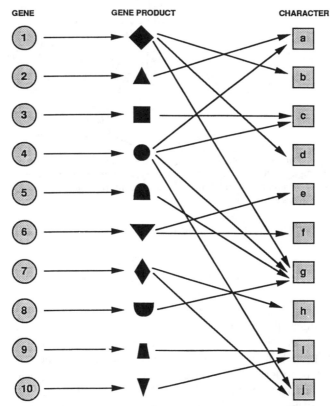

Figure 2-2 The effects of pleiotropy and polygeny. *Pleiotropy* is the effect where a single gene affects multiple phenotypic characters. *Polygeny* is the effect where a single phenotypic character is affected by multiple genes. (After Mayr, 1963, p. 265.)

describes them only on the average, and often assumes a one-gene/one-trait model of the relationship between the genotype and the phenotype. Although such a reductionist simplification is mathematically tractable and has been extensively analyzed (e.g., Fisher, 1930), it is not broadly useful in predicting or assessing the adaptation of behaviors. Such efforts have been labeled "bean-bag genetics" (Mayr, 1959, 1963, p. 263, 1988, p. 149; cf. Haldane, 1964). Bonner (1988, p. 24) suggested "evolution can no longer be looked at solely as changes in gene frequencies within populations."

The one-gene/one-trait model of evolutionary genetics is an over-simplification. Naturally evolved systems are extensively pleiotropic and highly polygenic. The importance of these profound genetic interactions on evolved structures has long been argued by Wright (1931), Simpson (1949, p. 224), Dobzhansky (1970), Stanley (1975), Mayr (1959, 1963, 1982, 1988), Dawkins (1986, p. 296), Hartl and Clark (1989, pp. 7, 149, 430, 501, 503–504), and others. Further, selection acts on collections of interactive phenotypic traits, not on singular traits in isolation. The appropriateness of an organism's holistic functional behavior in light of the physics of its environment is the sole quality that is optimized through selection.

The fundamental characteristics of diverse environments are often profoundly similar, and convergent evolution is the general outcome. The compound image-forming eye has been invented at least three times during the course of evolution: in mollusks, arthropods, and vertebrates (Salvini-Plawen and Mayr, 1977, indicated that simple light-sensitive organs have been evolved 40 to 65 times independently). The most common ancestor of these three taxa could have occurred no later than 450 million years ago and more probably 550 to 600 million years ago (Atmar, personal communication; Marshall, personal communication). But in all three cases, a profoundly similar informational neurophysiology has been invented, although slightly differently realized. In vertebrates, the neural connections lay on top of the retinal surface. In mollusks, they occur behind the eye as they do in arthropods. But in all three, a network of collateralized synaptic connections occur in a series of amacrine, horizontal, and bipolar cells that generate edge enhancement, motion detection, and modulatable light sensitivities in strikingly similar manners (Atmar, 1979). The convergent functional behavior of these independently evolved constructions leads to the conclusion that the invention of such a functional structure is exceedingly probable. Although there are obvious differences in the implementation of the eye arrangements, their functional behavior is profoundly similar. The physical and historical constraints woven into the mechanisms and process of the construction of a cell appear to predestine the evolution of a particular image-forming eye (Atmar, 1979).

Similarly, the invention of flight has occurred repeatedly. Gliding has been invented independently in fish, amphibians, reptiles, and mammals. Self-sustained flight has evolved independently in pterosaurs, birds, mammals, and insects (Bonner, 1988, p. 169). Other forms of flight have been invented in insects, arachnids, and plants. Young spiders, in which the young are transported over great distances by means of a parachute produced from silken thread, have been documented to drift over 400 miles (Sheppard, 1975). Plants similarly use airborne transportation to facilitate the dispersal of their seeds. In some cases, seeds have evolved an airfoil (e.g., *Zanonia macrocarpa;* Fogel, personal communication). The underlying genetic structures for these independently evolved phenotypic characteristics are diverse, yet the functionalities are notably similar.

Even ethologies are often profoundly convergent. During displays of aggression, animals as varied as Australian crayfish, baboon spiders, toads, wolves, and black-headed gulls all assume postures that make them appear larger to their adversaries. These behaviors include the raising of body hair, standing on hind legs in an upright manner, and spreading of the forelegs or claws (Keeton, 1980, pp. 517–518). Bonner (1988, p. 64; 1993, pp. 39–59) suggested that something as fundamental as multicellularity must also have been independently invented multiple times throughout the course of evolutionary history.

Pleiotropy and polygeny preclude the possibility of singular genes for these complex effects. While it is possible to alter specific traits through single gene mutations, this does not indicate a gene "for" that particular trait. By analogy, it is possible under the right circumstances to change a single card in a poker hand and to alter a flush (five cards of one suit) into a straight (five cards in ascending order). But that does not indicate that the single changed card is a "card for a straight." There are no individual genes for specific behaviors, such as "making oneself appear big," in natural evolved systems.

2.3   IMPLICATIONS OF WRIGHT'S ADAPTIVE TOPOGRAPHY: OPTIMIZATION IS EXTENSIVE YET INCOMPLETE

Wright (1932) offered the concept of the adaptive topography to describe the fitness of individuals and to serve to illustrate the manner in which evolution may proceed into novel adaptive zones. This concept has since become central to evolutionary theory. Individual (or populational) genotypes map into their respective phenotypes, which are in turn mapped onto the adaptive topography (Figure 2-3). Each peak corresponds with a fitness-optimized phenotype and, by consequence, one or more optimized genotypes. Evolution probabilistically proceeds up the slopes of the topography toward peaks as selection culls inappropriate phenotypic variants. But no natural topography is stable. Rather, the topography changes shape over time as a function of extrinsic environmental dynamics and organism-environment interactions. As the topog-

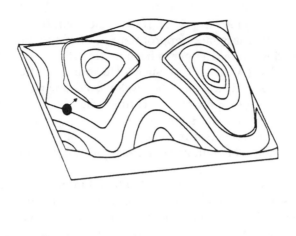

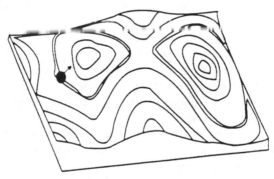

Figure 2-3 Evolution on an inverted adaptive topography. A landscape is abstracted to represent the fitness of alternative phenotypes and, as a consequence, alternative genotypes. Rather than viewing individuals or populations as maximizing fitness and thereby climbing peaks on the landscape, a more intuitive perspective may be obtained by inverting the topography. Populations proceed down the slopes of the topography toward valleys of minimal predictive error.

raphy shifts, the allelic balance in the population changes as the result of selection on the distribution of phenotypes.

The adaptive topography is a virtually continuous function because the range of phenotypic expression may also be regarded as virtually continuous (Hartl and Clark, 1989, p. 7). Boundary conditions, however, define the limits of possible evolutionary changes. Although not all genetic changes are possible (or are of infinitesimally small probability), the range of phenotypic expression can vary widely due to the multiple, nonlinear, and additive effects of gene products. Complex, indistinguishably different behaviors may be coded for by very different genetic programs. This indistinguishability, to a great degree, makes the structure of the genetic program generally irrelevant. Selection directly acts only on the phenotype.

Atmar (1979), Templeton (1982), Raven and Johnson (1986, p. 400), and others suggested that it is more appropriate to view the adaptive landscape from an inverted position. The peaks become troughs, "minimized prediction error entropy wells" (Atmar, 1979). Such a viewpoint is intuitively appealing. Searching for peaks depicts evolution as a slowly advancing, tedious, uncertain process. Moreover, there appears to be certain fragility to an evolving phyletic line; an optimized population might be expected to quickly fall off the peak under slight perturbations. But the inverted topography leaves an altogether different impression. Populations advance rapidly, falling down the walls of the error troughs.

Selection pressures are never wholly abated in nature. Indeed, the more normal circumstance is that selection pressures remain generally intense and drive phenotypes as close to an optimum as possible, given the initial and boundary conditions. "The picture emerging of natural selection at the level of the individual . . . is one of exuberance, power, and a potential for quickness" (Wilson, 1992, p. 92). But the environment is not static. It changes, and species' optimizations lag behind those changes, evolving toward new optima on the adaptive topography. It is impossible for a species to be perfectly adapted to its environment because of this flux in organism-environment interaction. Moreover, phenotypic characteristics take on a historical quality as adaptations to past environments may serve to restrict the potential for advancing into novel adaptive zones. While some phenotypic characteristics are deeply ingrained in the developmental sequence of the organism, other adaptations are maintained by a constant pressure of selection, without which they would degenerate due to random genetic drift.

All populations continually undergo selection for adaptations to the current environment. "Evolution is opportunistic, and natural selection makes use of whatever variation it encounters" (Mayr, 1988, p. 153). "Natural selection does not work like an engineer. It works like a tinkerer" (Jacob, 1977). The "baggage" of selection for adaptations to past environments imposes constraints on available current behaviors. Evolution simultaneously leads both to "organs of extreme perfection" (Darwin, 1859) and to global suboptimality. Functionally critical behaviors may be expected to be extremely well optimized. Behaviors that are less critical may be expected to vary to a greater degree.

2.4   THE EVOLUTION OF COMPLEXITY: MINIMIZING SURPRISE

A general trend of evolution is the generation of organisms of increasing complexity. Simon (1962) defined complexity as that situation when "given the properties of the parts and the laws of their interaction, it is not a trivial matter to infer the properties of the whole," while Bonner (1988, p. 98) has described the complexity of a system not only in terms of the number of interacting parts, but also in terms of their differences in structure and function. These differences often lead to divisions of labor in biological systems (e.g., number of cell types). Bonner (1988) emphasized these sorts of interactions, along with hierarchies of size, as being most fundamentally important. But there are special problems in examining systems of such extraordinary complexity. "Each organic system is so rich in feedbacks, homeostatic devices, and potential multiple pathways that a complete description is quite impossible. Thus, a prediction of its productions is also impossible. Furthermore, the analysis of such a system would require its destruction and thus preclude the completion of the analysis" (Mayr, 1982, p. 49). In examining the fossil record and the estimated sequence of organisms, the "inescapable conclusion [is] that the earliest forms were simple and small, and that as evolution has progressed there appeared increasingly complex and larger plants and animals" (Bonner, 1988, p. 161).

Increasing complexity is required to advance the quality of prediction of environmental events. The environment can be viewed as an nth-order, non stationary stochastic process where the previous n symbols (e.g., the motion of the sun across the sky) influence or are associated with the probability of the occurrence of the next symbol. Should the correlation of certain strings of symbols be high, then it is to be expected that the evolving phyletic line will come to optimally predict ever longer symbol strings; uncorrelated sources will tend to be ignored in favor of more certain portents of future events. If the universe were completely uncorrelated in its symbol strings of increasing length, optimal prediction would degenerate to just the most likely symbol. But the universe is not uncorrelated, and the languagelike characteristics of phenotypic behaviors incorporated into an evolving phyletic line are readily apparent.

There is enormous complexity in living systems, in a cell, an organ, an individual, a population, a species, or an ecosystem. "Living systems are characterized by elaborate feedback mechanisms unknown in their precision and complexity in any inanimate system" (Mayr, 1982, p. 53). Complexity in the phenotype is not random, but rather is well organized. Such complexities have been documented at various levels ranging from multicellularity to animal behavior (Bonner, 1988). Complexity does not persist for its own sake but rather is maintained as "entities at one level are compounded into new entities at the next higher level—cells into tissues, tissues into organs, and organs into functional systems" (Mayr, 1988, p. 14). Increased complexity allows for a greater range of potential behaviors and thus provides a competitive advantage.

2.5 SEXUAL REPRODUCTION

The term *sex* derives from the Latin *secare,* to cut or divide something that was originally whole. The exchange of genetic information is fundamental to the cutting and dividing of DNA that takes place during sexual reproduction. This exchange takes many forms: bacterial conjugation, transduction via viral transport, and the more familiar recombination that occurs in plants and animals. But sex requires finding a mate, either by happenstance, active search, or attraction. Further, the behaviors that are associated with attracting mates often make individuals more vulnerable to predators. Yet the benefits of sexuality must outweigh these costs, for the great majority of known species have a sexual phase in their life cycle.[1]

Diploid organisms have two copies of each chromosome in every somatic cell, with one copy being contributed from each parent. The total number of chromosomes in the cells of most eukaryotes (organisms with fully nucleated cells) varies between 10 and 50 but can exceed 1,000 in some polyploid plants. During recombination, haploid (nonredundant) gametes are fused to form new diploid cells.

A reduction division is then required for this infusion to generate a stable number of chromosomes in the sexually reproducing lineage. This division is termed *meiosis*. In this process, each chromosome from both parents is duplicated. A crossing-over between homologous chromosomes may occur in which segments of the chromosome are exchanged. The rate of crossing-over varies widely from species to species and between sexes. Male fruit flies (*Drosophila* sp.) exhibit no crossing-over at all, while in mammals the rate of crossing-over is about 30 percent higher in females (Gould and Gould, 1989, p. 34). During meiosis, a reduction division occurs in which one of the two homologous chromosomes is sorted into each cell. If crossing-over does take place, it occurs predominantly during this first meiotic division. The chromatids then undergo a second reduction division such that each gamete receives a potentially hybrid combination of each pair of homologous chromosomes from each parent.

In contrast, sexual reproduction operates in a different manner. Sexual reproduction offers a significant ability for a species to generate genetic diversity and, as a consequence, phenotypic diversity. The overt functional advantage of sexual recombination to an evolving species is the significantly increased rate of exploration of the genotypic/phenotypic state space, as compared with clonal parthenogenesis, especially in changing environments. Recombination will ultimately tend to expose a wide variety of individual genotypes to various environmental conditions. The number of such genetic combinations is, for all practical purposes, unlimited. It is presumed that each offspring receives each chromosome from each parent in a random manner. In

---

[1]Wilson (1992, p. 134) places the number of *known* species at 1,413,000, with 751,000 of these being insects, 248,400 being higher plants, and 281,000 being other animals. Virtually all these have a sexual phase in their life cycle. Yet Wilson also comments that vast arrays of species remain to be found in bacteria, fungi, and other groups, and he estimates that there are somewhere between 10 and 100 million species on the planet. The majority of species are yet undiscovered. How common sex is for the undiscovered species remains to be observed.

humans, 2^{23} distinct zygotes can be produced from only two parents by simply considering the possible combinations.

Errors are inevitable during this reshuffling of information. The species' genetic program carried in every cell of every individual's body is not recently evolved, nor is it simple. Rather, it is ancient at the extreme and enormously complex. Maintaining the fidelity of the genetic information is critical. The enzyme complex that transcribes the chromosome appears to error check the nucleotide chain being replicated, replacing the appropriate base to the complementary strand as replication proceeds. Further error checking ("proofreading") is accomplished by checking for mismatches. But there are also many environmental sources of mutational variation. These may result in deletions, substitutions, insertions, and inversions of sections of DNA, all forms of entropic informational decay. Mutations that affect both strands of DNA are difficult to repair because a reference copy is generally used as a guide. Such randomization of genetic information will quickly fail to produce competitive phenotypes. A lack of competitive phenotypes implies severe costs to the species unless information maintenance mechanisms are invoked to maintain the functionality of the genotype (Atmar, 1991).

Three potential advantages of recombination have been identified (Crow, 1988): (1) greater efficiency for adjusting to a changing environment, (2) bringing together independently created beneficial mutations, and (3) removing deleterious mutations. The relative importance of these factors remains under debate, although there appears to be broad agreement that removing deleterious mutations is a fundamental attribute of recombinatory mechanisms (e.g., Crow, 1988; Bernstein et al., 1988; Bell, 1988). Bernstein et al. (1988) have argued that endogenous repair was the original function for which recombination served. Bell (1988) suggested that a second method for repair is to subject each individual to stringent tests of performance and to eliminate those that fail to attain a tolerable degree of achievement (exogenous repair). Under this view, recombination serves to expose genetic defects to the culling force of selection. There is less agreement that recombination is maintained because it can bring together favorable mutations (Crow, 1988), and certainly the relevance of recombination for adjusting to changing environments (Bell, 1988) must depend on the degree to which the organism's environment is in flux.

2.6   SEXUAL SELECTION

Sexual reproduction allows for *sexual selection*. This term refers to the process by which males and females of a species determine or "select" their mates. These processes often involve prolonged demonstrations of vigor between members of the species, typically the males. Male-male competition has been viewed as an important source of sexual selection since Darwin (1859). He concluded that the result of these encounters was not generally death to the vanquished, but "few or no offspring" (Gould and Gould, 1989, p. 86). A genetic death is equally as real as a physiological death in purging a defect from the germline.

Males in many species have evolved specific attributes that appear designed solely for male-male competition. Antlers, for example, are used primarily in intraspecies competition but otherwise appear poorly constructed for predator defense. The enlarged claw of male crabs (especially fiddler crabs, in which the pincer may constitute one-third of the crab's total body weight) is similarly observed only in mature males and is not required for survival. Rather, the claw is used in wrestling matches between males and for attracting females.

Aggressive behavior is typical of virtually all males and can be triggered by specific stimuli. In experiments with stickleback fish, Tinbergen (1951, pp. 27–28) indicated that males are more aggressive toward decoys with a red coloration but no visual resemblance to a fish than towards anatomically correct, colorless decoys. Similar results have been described by Lack (1946, pp. 144–147) in which male robins threatened a pile of red feathers more readily than a stuffed young male that lacked a red breast. Male sexual maturation is often accompanied by these sex-specific characters that incite aggressive behavior in other males.

There are numerous examples of male-male competition in a great variety of diverse species (Gould and Gould, 1989, pp. 85–265). A male stag beetle has overdeveloped horns and an enlarged mandible that appear specialized for intraspecies fighting. Combat between male stalk-eyed flies in Australia includes a ritual in which the males line up eye-to-eye and apparently determine which individual has a wider-set pair of eyes. Male salmon returning from the sea to spawn develop a hooked jaw that aids in fights with other males. As the males die shortly after spawning, the problems in feeding caused by the hooked jaw are of no consequence. Male elephant seals, due to the very limited breeding ground, are under intense sexual selection, and high mortality rates are not uncommon.

Demonstrations of male vigor appear to be a principal component of the evolutionary process that exposes and expurgates significant genetic errors from germline DNA. Those males that prove themselves "fragile" through prolonged competition do not mate. Males have often become a sacrificial component of a reproductive subpopulation. Sexual selection may be viewed as operating as a gene defect filter by continually testing the male germline for vigor (Maynard Smith, 1988, p. 177; Atmar, 1991), and such male-male competition serves to cull genetic defects from the breeding deme prior to reproduction.

2.7 ASSESSING THE BENEFICIARY OF EVOLUTIONARY OPTIMIZATION

Evolution is an optimization process (Mayr, 1988, p. 105). Selection statistically eliminates suboptimal phenotypes. Although the purposes of many of the evolved behaviors are transparent, the underlying reason for, and the ultimate beneficiary of, these behaviors has remained an open question that has undergone considerable debate. There are at least three commonly stated and plausible hypotheses: The benefits of optimization accrue to (1) the individual gene, (2) the individual's genome, or (3) the reproducing population.

If evolution is expected to lead to perfected genes, the nature of the gene must be explained. The early work of Morgan with *Drosophila* led to the conception of the gene as the smallest unit of recombination. "Two characters were regarded as determined by different genes if they could be recombined, and as determined by the same gene if they could not be recombined" (Keeton, 1980, p. 656). Early on, genes were conceptualized as particles linearly arranged on chromosomes. Recombination was explained by a crossing-over mechanism that broke the chromosomes apart at points between these ordered particles and reassembled appropriate chromosomal segments.

Currently, a gene is commonly defined as the genetic information necessary to promote the synthesis of one polypeptide chain (Raven and Johnson, 1986, p. G-12). But it is also known that in eukaryotes, a gene and any given segment of DNA are rarely colinear. "Any 'single' gene, in the sense of a single continuously read passage of DNA text, is not stored in one place" (Dawkins, 1986, p. 174). Because of the complex and interwoven informational mappings, Lewin (1983) suggested that a gene should be redefined as one polypeptide/one gene, reversing the previous definition, allowing the functional product to define the gene. Because selection acts immediately only on the phenotype, and only by consequence on the underlying genotype, it would seem far more prudent to emphasize the functionality of the cohesive genotype rather than its apparent structure.

The danger lies in the promotion of an atomistic "beanbag" perspective of evolution that has led to proposals of "selfish genes" (Dawkins, 1976, 1982). But there is an obvious problem with the proposition that individual genes are the units of selection: Individual genes cannot be assigned a meaningful fitness value in isolation. The phenotypic impact of any single gene varies as a pleiotropic and polygenic function of all other genes and as a function of the environment. Selection cannot take place in the absence of competition, and only through competition is relative fitness assessed.

Only a small percentage of the eukaryotic genotype actually codes for expressible behavior (Lewin, 1986; Mayr, 1988). The great majority of human DNA is untranslated (Lewin, 1986; Brunk, 1991). It is somewhat unreasonable, therefore, to speak of evolution producing a set of optimized genes when much of the genome is "junk" (Lewin, 1986). Moreover, assigning absolute fitness to genes leads "to an overvaluation of additive gene effects . . . [and] invariably [leads] to unrealistic results" (Mayr, 1982, p. 41). Mayr (1982, p. 400) has criticized such an explanation as meaningless because "it tells us nothing about the multiplication of species nor, more broadly, about the origin of organic diversity."

A second alternative would assert that evolution operates such that benefit accrues to the individual or its genome. But if the individual is the beneficiary of optimization, then that benefit does not persist long, as an individual's life span is minute in any evolutionary sense and its genetic legacy is rapidly washed away in an infinite sequence of sexual divisions and fusions. If evolution were truly to act to optimize individual genotypes, then far greater rates of clonal parthenogenesis must be expected than are observed. Once the best possible genotype was found, the optimum strategy under this physics would be to copy it endlessly. But the overwhelming majority of species are sexual. Further, if the individual acts solely for its own benefit, then many observed altruistic behaviors appear paradoxical.

Hamilton (1964), and earlier Haldane (1932), suggested that an explanation of apparent altruism could come in the form of *kin selection*. This argument asserts that selection operates on the level of individuals so as to increase the frequency of genetic representation of an individual in future generations. Brothers and sisters are often viewed as sharing half of their genes, on the average, so they would be more apt to act altruistically toward each other than toward more distant relations from the same species or toward completely distinct species. Sacrificing oneself for three brothers is viewed as a "selfish" act of propagating more copies of one's own genes into future generations.

This explanation has been criticized by Sober (1992) because, due to the effects of selection, siblings almost never share only half of their genes. "A population in which natural selection has destroyed lots of variation may be such that all individuals are very similar to each other, and full sibs are even more so (Dawkins, 1979)" (Sober, 1992). Nevertheless, this mechanism (a relatedness of $r = \frac{1}{2}$ between siblings or otherwise as appropriate) is used to explain the apparent altruism of the members of sexually sterile castes, for instance in social insects (bees, ants) in which these workers exist solely to serve the hive or colony and never to reproduce themselves (Hartl and Clark, 1989, p. 563). Kin selection is seen as maximizing *inclusive fitness,* the probability that a specific genetic representation will be copied into future generations.

Despite wide-ranging support for this idea (Bonner and May, 1981; Dawkins, 1982, pp. 193–194; and others), the argument is open to criticism. Selection is now no longer truly acting on the individual as a whole, but rather on fractions of the genotype pool. Specifically, selection is viewed as acting to increase the chances that the largest possible fraction of an individual genotype will be multiplied and will increase in frequency (Hoffman, 1989, p. 138). The individual is dissolved into fractions of genetic representation that are supposed to be acting in support of their own reproduction. As a unit of selection, the individual "dissolves into vagueness" (Hoffman, 1989, p. 140). The argument ultimately reduces to viewing individual genes as independent units affecting the phenotype of individuals. But as has been well documented by Mayr (1970, pp. 162–185), there is a unity to the genotype. "No gene has a fixed selective value; the same gene may confer high fitness on one genetic background and be virtually lethal on another" (Mayr, 1970, p. 184). That is, the fraction of similarity between two individual genotypes may be absolutely irrelevant. Two individuals may differ in but a single gene out of thousands or tens of thousands, yet this difference may be associated with a lethal defect. It is insufficient simply to maximize the fraction of genetic representation in future generations. The fraction not represented is absolutely crucial to the realized phenotypic effect.

If evolution is to be viewed as a process such that advantage accrues to individuals it is reasonable to ask, "Is it . . . evolutionarily to each parent's advantage to form a monogamous pair bond and make a large investment in the young?" (Keeton, 1980, p. 544). The question implicitly assumes that some benefit for creating offspring accrues to the parents. But that benefit must be identified. The argument presupposes that each parent has an interest in leaving offspring, typically the greatest possible number of offspring. But individuals must compete for survival with other members of the population, and the most intense competition comes from members of the same species that are, essentially, behaviorally identical. An over-abundance of the species is a real threat to all members of the species because available resources are ultimately limited. The fundamental reproductive drive becomes an explanation after the fact if individuals are to be viewed as the ultimate beneficiaries of evolutionary invention. "Animals sometimes behave in a way that makes no sense from the point of view of maximizing individual fitness" (Hartl and Clark, 1989, p. 560).

Evolution is a learning process. Phylogenetic learning (i.e., learning arising from within the species) is the most commonly exhibited and most ancient form of intelligence. Through mutation and selection there is an obvious optimization of behavior. As described in Chapter 1, Atmar (1976), following Weiner (1961), proposed phylogenetic, ontogenetic, and sociogenetic learning to be equivalent processes, each containing a quantum unit of mutability and a reservoir of learned behavior. In phylogenetic learning, the quantum unit of mutability is a single nucleotide base pair. The reservoir of learned behavior is the species' genome. From such a perspective, the unit of selection is the individual, but the ultimate beneficiary of evolutionary invention is the species, or at least the reproducing population.

The notion that individuals forgo certain possible benefits "for the good of the species" has become associated with the term *group selection*, which was strongly advocated by Wynne-Edwards (1962). This argument asserts that competition for survival occurs between entire groups of individuals and that groups replace individuals as the unit of selection while speciation replaces individual reproduction.

Group selectionist explanations have fallen out of favor with many evolutionary biologists following counterarguments by Williams (1966) and Maynard Smith (1976). Even Mayr (1970, p. 115), who has commented that reproductive populations benefit from selection, remarked that "all attempts to establish group selection as a significant evolutionary process are unconvincing." Hoffman (1989, p. 137) relates a good example, and even though the following quote is lengthy, it is worth repeating:

> John Maynard Smith (1976) made a very simple but powerful argument to this effect. Consider a population the members of which have been conditioned by group selection to forgo some potential advantages to their individual offspring and to cooperate with each other for the sake of their common welfare. If such a population is invaded by an immigrant that still acts exclusively for the sake of its own offspring—to increase as much as possible the fraction of the population its progeny will constitute in future generations—it shall certainly do better in reproducing within this population than its more altruistic members. Provided therefore that these behaviors have a genetic background and are inheritable, an assumption shared by the theory of group selection, the selfish individuals will soon outcompete the altruists in the population. In the nonhuman world, altruistic behaviors should therefore occur at best temporarily, until selfishness takes over again.
>
> Thus, group selection cannot be the solution to the dilemma that the phenomena of apparent altruism poses to the genetic theory of evolutionary forces.

To escape the apparent paradox that altruistic behavior poses, many biologists have adopted Hamilton's (1964) kin selection.

But the argument by Maynard Smith (1976) has an implicit assumption that restricts its application. The assumption is that the conditions of the experiment remain constant. That is, altruistic (and therefore mutually cooperative) individuals will continue to adopt this policy, even when encountering the selfish individuals. Under such conditions, Maynard Smith's (1976) conclusions are correct: Selfish individuals would certainly outcompete altruistic individuals and would quickly come to dominate the population. If there is no mechanism for the altruists and selfish actors to identify other members of the population as either altruists or selfish actors, and if behavior is genetically preprogrammed into each individual, then intense competition is the likely result (e.g., Lysenko's clusters of oak trees; see Hoffman, 1989, p. 137).

But suppose that the individuals in the population evolve mechanisms that allow them to recognize other individuals as either altruistic or selfish. Further suppose that individuals can vary their policies based on these recognitions. There are several expected results under these new conditions. The first is that altruists would not only continue to cooperate amongst themselves (enhancing the benefits to the group), but they would also ostracize selfish individuals. Another likely outcome is that some of the altruists would instead adopt selfish policies and, at least for some time, continue to be recognized for their previous altruistic behavior and thereby take advantage of other altruists by fooling them into cooperating. Further, some selfish individuals would adopt behaviors associated with altruists, thereby fooling other altruists into cooperating only to take advantage of them at a later time. Whether or not mutual cooperation or selfishness rise to dominate the population then becomes a question of evolutionary dynamics: How quickly can recognition and deception mechanisms be evolved? In fact, there may be no stable result. But the observed evidence indicates that recognition mechanisms are pervasive in evolved organisms and that behavior does vary in light of the recognition of expected encounters, so the easy analysis of Maynard Smith (1976) simply does not apply to the real circumstance and may be dismissed.

What cannot be so easily dismissed is the idea of groups replacing individuals as the unit of selection. It is clear that individuals compete for existence. When a pack of wolves bring down a deer, there can be no doubt that selection acted against the individual deer. The term *group selection* serves to discredit the idea that groups are the ultimate beneficiaries of selection. There is no logical necessity for the unit of selection to correspond directly to the unit for which adaptations serve to benefit (cf. Williams, 1966, p. 92; Dawkins, 1982, p. 81; Maynard Smith, 1988, pp. 54, 186). Instead, the reproductive population can be seen as the beneficiary of selection acting against individuals, culling inappropriate (i.e., self-defeating) individual behaviors.

Previously paradoxical or anomalous behaviors are more easily explained when optimization is viewed as accruing to the reproducing population. Sexual recombination serves to verify allelic combinations rapidly and exhaustively. Those that are nonviable are quickly sieved from the population. Sexual selection, and more specifically male-male contest, serves as a secondary genetic error expurgation mechanism; the viability of the species' germline is enhanced through the aggressive competition between sexually mature males (Atmar, 1991). It is not unreasonable to expect that individuals of certain populations will become sacrificial, never mate, and only serve to protect those individuals that reproduce.

Just as the individual's fitness is a function of its entire cohesive genetic composition—and no genetic structure can be measured in isolation—the population's fitness is a function of the entire cohesive set of behaviors of its individuals. When the appropriateness of individuals' behavior is viewed outside of the context of the species, behaviors often appear inexplicable or even irrational. These observations, however, can generally be reconciled when attention is focused on the reproductive population as a whole.

2.8   CHALLENGES TO NEO-DARWINISM

The neo-Darwinian view of evolution has not gone unchallenged. The question of whether microevolution (evolution within populations and species) can account for evolution on a larger scale has been debated at length. Goldschmidt (1940) believed that the genetical forces postulated by neo-Darwinists were not sufficiently powerful to create new species and argued that macromutations ("hopeful monsters") are requisite to speciation. In distinct contrast, Simpson (1944) instead argued that all historical biological phenomena are brought about by the same evolutionary forces acting at all levels; Simpson's perspective has prevailed for more than four decades. Recently, there have once again been claims that the traditional neo-Darwinian view does not explain the observed evidence of evolution, and alternative hypotheses have been proposed.

2.8.1   Neutral Mutations and the Neo-Darwinian Paradigm

Kimura (1983) proposed that most changes in gene frequencies (a common definition of evolution) are of no selective value, but are in fact neutral. It had previously been believed that "there is no evidence for the existence of genes that remain selectively 'neutral' " (Mayr, 1963, p. 159). Recent research has in fact established that many changes at the molecular level are entirely neutral. Those who were predisposed to believe that genes are the targets of selection were therefore left an opening to argue less emphasis on natural selection. Such models have been criticized as over-emphasizing the importance of individual genes (Mayr, 1982, p. 102; and others). If the individual as a whole is taken as the target of selection, then numerous neutral genes, or even potentially harmful genes, can be carried as "hitchhikers," as long as the individuals survive selection due to other expressed characteristics. Although the neutral theory was originally thought to be in conflict with neo-Darwinism, in fact, the mere existence of neutral mutations does not violate the neo-Darwinian paradigm (Brunk, 1991). At best, it only offers a mechanism of phenotypic variation that is not directly attributable to the intense selection pressures commonly associated with natural selection.

2.8.2   Punctuated Equilibrium

Punctuated equilibrium, offered by Eldredge and Gould (1972), is perhaps the most well-known novel hypothesis. They proposed that the fossil record accurately reflects periods of extremely (geologically) rapid morphological change associated with speciation, followed by prolonged periods of stasis. This is in contrast to the hypothesis of *phyletic gradualism,* which asserts that morphological transformations are the result of the accumulation of small changes occurring over long periods of time, with the apparent gaps in the fossil record being the result of incomplete data. Punctuated equilibrium has received various interpretations, ranging from it being a "minor gloss" on neo-Darwinism (Dawkins, 1986, p. 250) to it being a "revolt against the principle of continuity" (Rieppel, 1987).

Hoffman (1989, p. 100) characterized two versions of punctuated equilibrium: *weak* and *strong*. The weak version asserts that the evolution of species does not proceed for long in the same direction, nor does so with a constant rate, but rather varies in both direction and rate. This is no challenge to neo-Darwinism, as the neo-Darwinian paradigm certainly predicts a variation in the rates and directions of evolution as populations enter novel adaptive zones. Indeed, these variations have been documented by Simpson (1944, 1953). The strong version of punctuated equilibrium, which may be viewed as somewhat extreme, argues that gradual phenotypic change is negligible in the evolution of species, with the gaps between successive species being dominant. Hoffman (1989, p. 105) described the problems of validating this version with evidence in the fossil record, which is not at all complete and is subject to enormous variation. Moreover, there is evidence of a continuous range of evolutionary rates, rather than the distinctly bimodal distribution that would be expected under the strong version (Gingerich, 1983).

Some advocates of punctuated equilibrium have proposed that there are fundamental differences between *microevolution* (within a species) and *macroevolution* (generation of new species) (Stanley, 1979). Macroevolution has been described as requiring macromutations, mutations that introduce large effects in the phenotype, with the result being "rectangular evolution" (Stanley, 1975). Gould (1980) apparently supported a "potential saltational origin for the essential features of key adaptations. Why may we not imagine that gill arch bones of an ancestral agnathan moved forward in one step to surround the mouth and form proto-jaws?" Yet more recently, Gould and Eldredge (1993) indicated that the interpretations of punctuated equilibrium as saltational theory are "misunderstandings . . . that misread geological abruptness as true suddenness" and do not support such a view.

Although attention was originally focused on the apparent rapid change associated with punctuations in the paleontological record, it has now turned to the apparent stasis of many species (Gould and Eldredge, 1993). The argument asserts that because species often maintain stability even during times of dramatic climatic change (e.g., glacial cycling), stasis must be viewed as an active phenomenon rather than merely a passive response to unaltered environments. Mayr (1988, p. 468) admitted that "I agree with Gould that the frequency of stasis in fossil species revealed by recent analysis was unexpected by most evolutionary biologists," but also argued that prolonged stasis is not in contention with neo-Darwinism, for any major morphological restructuring would require a drastic interference to the cohesion of the genotype and would be selected against quite vigorously (Mayr, 1988, p. 434). Punctuated equilibrium as offered in Gould and Eldredge (1993) does not present a challenge to the basic structure of neo-Darwinism. The patterns of prolonged stasis and rapid morphological change are consistent with neo-Darwinism, even if they are not fully anticipated by it.

2.9   SUMMARY

The notion of Darwinian evolution is essentially quite simple. In fact, Gould (1977, p. 11) and Mayr (1988, p. 117) have described the principle of evolution as "simplicity itself," and yet this apparently obvious principle was resisted for three-quarters of a century after being proposed by Darwin and Wallace. Mayr (1988, p. 117) speculated that the reasons for the resistance were: (1) Darwin was unable to explain the source of phenotypic variations, (2) Darwin was unable to prove the occurrence of natural selection, (3) a probabilistic theory was unpalatable in a time when the universe was viewed as being deterministic, and (4) scientific thought was essentialist, viewing the population solely in terms of an average behavior and regarding variation as nothing but errors around the mean values (Mayr, 1982, p. 47). The selection that is so critical to Darwin's evolution is untenable in the absence of a populational view.

Evolution most aptly refers to changes in adaptation and diversity, and not directly to changes in gene frequencies. Gene frequencies persist only as a result of selection on the phenotype. The genetic variation of individuals is largely a chance phenomenon, primarily a product of recombination and "ultimately of mutation" (Mayr, 1988, p. 532). Selection is probabilistic, and its primary target is the individual. Sexual reproduction allows for the rapid phenotypic exploration of an underlying genetic theme. It also allows for additional error correction mechanisms (e.g., sexual selection). Nonetheless, the specific mechanisms involved in recombination are unimportant from an optimization perspective, except for the manner in which they affect individual phenotypes.

The individual is the target of selection, but the species, or minimally, the reproducing population, is the ultimate genetic reservoir of all learned behavior. It is an evolutionary element that can have a measurable fitness and possess a degree of permanence. The notion that the species is the ultimate beneficiary of evolutionary optimization has been proposed by Mayr (1963, 1988), Conrad (1990), Atmar (1991), and others. But Hecht and Hoffman (1986) and many others appear to view the process of optimization accruing to the species as an alternative to individual selection, rather than the result of individual selection. Yet if all individuals are viewed as acting in their own best interest, many of the observed behaviors appear paradoxical. Arguments identifying the teleonomic (purpose-driven) (Mayr, 1988, p. 45) behavior of individuals as benefiting their respective species appear more appropriate and concise. Arguments that attribute such behaviors to a desire to pass along an individual's genetic structure may appear somewhat contrived.

Lindsay (1968) commented, "The theory of evolution is not typical of scientific theories. Rather than prescribing a specific process or formula it merely provides a plausibility argument involving a couple of general principles and gains broad acceptance by virtue of the lack of plausible alternatives." But the neo-Darwinian paradigm, the modern synthesis of evolutionary thought, lays out a well-defined process that extends beyond a mere plausibility argument. Individuals reproduce sufficiently well so as to expand to fill the available resource space. Errors during genetic replication result in individual phenotypic variation. Furthermore, due to the nature of the genetic program, these phenotypic variations are heritable. Competition results from the inevitable excess population and leads to stochastic selection against those individuals that are less well adapted to the present environment. Randomness is the propellant of the evolutionary process; the adaptive topography is its primary determinant. Darwinian evolution is no more than the inevitable consequence of competing information-reproducing systems operating within a finite arena in a positively entropic universe (Atmar, 1992).

REFERENCES

Atmar, J. W. (1976). "Speculation on the Evolution of Intelligence and Its Possible Realization in Machine Form." Sc.D. diss., New Mexico State University, Las Cruces.

Atmar, J. W. (1979). "The Inevitability of Evolutionary Invention." Unpublished manuscript.

Atmar, W. (1986). Personal Communication, AICS Research, Inc.

Atmar, W. (1991). "On the Role of Males," *Animal Behaviour,* Vol. 41, pp. 195–205.

Atmar, W. (1992). "On the Rules and Nature of Simulated Evolutionary Programming." In *Proc. of the First Ann. Conf. on Evolutionary Programming,* edited by D. B. Fogel and W. Atmar. La Jolla, CA: Evolutionary Programming Society, pp. 17–26.

Atmar, W. (1994). "Notes on the Simulation of Evolution," *IEEE Trans. on Neural Networks,* Vol. 5:1, pp. 130–148.

Bell, G. (1988). "Uniformity and Diversity in the Evolution of Sex." In *The Evolution of Sex: An Examination of Current Ideas,* edited by R. E. Michod and B. R. Levin. Sunderland, MA: Sinauer, pp. 126–138.

Berstein, H., F. A. Hopf, and R. E. Michod (1988). "Is Meotic Recombination an Adaptation for Repairing DNA, Producing Genetic Variation, or Both?" In *The Evolution of Sex: An Examination of Current Ideas,* edited by R. E. Michod and B. R. Levin. Sunderland, MA: Sinauer, pp. 139–160.

Bonner, J. T. (1988). *The Evolution of Complexity by Means of Natural Selection.* Princeton, NJ: Princeton University Press.

Bonner, J. T. (1993). *Life Cycles: Reflections of an Evolutionary Biologist.* Princeton, NJ: Princeton University Press.

Bonner, J. T., and R. M. May (1981). Introduction to *The Descent of Man, and Selection in Relation to Sex,* by C. Darwin. Princeton, NJ: Princeton University Press (reprint).

Brunk, C. F. (1991). "Darwin in an Age of Molecular Revolution," *Contention,* Vol. 1:1, pp. 131–150.

Conrad, M. (1990). "The Geometry of Evolution," *BioSystems,* Vol. 24, pp. 61–81.

Crow, J. F. (1988). "The Importance of Recombination." In *The Evolution of Sex: An Examination of Current Ideas,* edited by R. E. Michod and B. R. Levin. Sunderland, MA: Sinauer, pp. 56–73.

Darwin, C. (1859). *On the Origin of Species by Means of Natural Selection or the Preservations of Favored Races in the Struggle for Life.* London: John Murray.

Dawkins, R. (1976). *The Selfish Gene.* Oxford: Clarendon Press.

Dawkins, R. (1979). "Twelve Misunderstandings of Kin Selection," *Zeitschrift für Tierpsychologie,* Vol. 51, pp. 184–200.

Dawkins, R. (1982). *The Extended Phenotype.* Oxford: Oxford University Press.

Dawkins, R. (1986). *The Blind Watchmaker.* Oxford: Clarendon Press.

Dobzhansky, T. (1937). *Genetics and the Origins of Species.* New York: Columbia University Press.

Dobzhansky, T. (1970). *Genetics of the Evolutionary Processes.* New York: Columbia University Press.

Dobzhansky, T., F. J. Ayala, G. L. Stebbins, and J. W. Valentine (1977). *Evolution.* San Francisco: W. H. Freeman.

Eldredge, N., and S. J. Gould (1972). "Punctuated Equilibria: An Alternative to Phyletic Gradualism." In *Models of Paleobiology,* edited by T. J. M. Schopf. San Francisco: W. H. Freeman, pp. 82–115.

Fisher, R. A. (1930). *The Genetical Theory of Natural Selection.* Oxford: Clarendon Press.

Fogel, G. B. (1993). Personal Communication, UCLA.

Gingerich, P. D. (1983). "Rates of Evolution: Effects of Time and Temporal Scaling," *Science,* Vol. 222, pp. 159–161.

Goldschmidt, R. (1940). *The Material Basis of Evolution.* New Haven, CT: Yale University Press.

Gould, J. L., and C. G. Gould (1989). *Sexual Selection.* New York: Scientific American Library.

Gould, S. J. (1977). *Ever Since Darwin: Reflections in Natural History.* New York: W. W. Norton.

Gould, S. J. (1980). "Is a New and General Theory of Evolution Emerging?" *Paleobiology,* Vol. 6, pp. 119–130.

Gould, S. J., and N. Eldredge (1993). "Punctuated Equilibrium Comes of Age," *Nature,* Vol. 366, November 18, pp. 223–227.

Haldane, J. B. S. (1932). *The Causes of Evolution.* London: Longman.

Haldane, J. B. S. (1964). "A Defense of Beanbag Genetics," *Perspectives in Biology and Medicine,* Vol. 7, pp. 343–359.

Hamilton, W. D. (1964). "The Genetical Evolution of Social Behavior: I and II," *J. Theor. Biol.,* Vol. 7, pp. 1–52.

Hartl, D. L., and A. G. Clark (1989). *Principles of Population Genetics,* 2nd ed. Sunderland, MA: Sinauer.

Hecht, M. K., and A. Hoffman (1986). "Why Not Neodarwinism? A Critique of Paleobiological Challenges," *Oxford Surv. Evol. Biol.,* Vol. 3, pp. 1–47.

Hoffman, A. (1989). *Arguments on Evolution: A Paleontologist's Perspective.* New York: Oxford University Press.

Huxley, J. S. (1942). *Evolution: The Modern Synthesis.* London: Allen and Unwin.

Huxley, J. (1963). "The Evolutionary Process," In *Evolution as a Process,* edited by J. Huxley, A. C. Hardy, and E. B. Ford. New York: Collier Books, pp. 9–33.

Jacob, F. (1977). "Evolution and Tinkering," *Science,* Vol. 196, pp. 1161–1166.

Keeton, W. T. (1900). *Biological Science,* 3rd ed. New York: W. W. Norton.

Kimura, M. (1983). *The Neutral Theory of Molecular Evolution.* Cambridge, England: Cambridge University Press.

Lack, D. L. (1946). *The Life of the Robin.* London: H. F. and G. Whiterby.

Lewin, B. (1983). *Genes.* New York: John Wiley.

Lewin, R. (1986). " 'Computer Genome' Is Full of Junk DNA," *Science,* Vol. 232, pp. 577–578.

Lewontin, R. C. (1974). *The Genetic Basis of Evolutionary Change.* New York: Columbia University Press.

Lindsay, R. K. (1968). "Artificial Evolution of Intelligence," *Contemp. Psychology,* Vol. 13:3, pp. 113–116.

Marshall, C. (1993). Personal Communication, UCLA.

Maynard Smith, J. (1976). "Group Selection," *Quart. Rev. Biol.,* Vol. 51, pp. 277–283.

Maynard Smith, J. (1988). *Did Darwin Get It Right?: Essays on Games, Sex, and Evolution.* New York: Chapman and Hall.

Mayr, E. (1942). *Systematics and the Origin of Species.* New York: Columbia University Press.

Mayr, E. (1959). "Where are We?" *Cold Spring Harbor Symp. Quant. Biol.,* Vol. 24, pp. 409–440.

Mayr, E. (1960). "The Evolution of Life." Panel in *Evolution after Darwin: Issues in Evolution,* Vol. 3, edited by S. Tax and C. Callender. Chicago: University of Chicago Press.

Mayr, E. (1963). *Animal Species and Evolution.* Cambridge, MA: Belknap Press.

Mayr, E. (1970). *Populations, Species, and Evolution.* Cambridge, MA: Belknap Press.

Mayr, E. (1982). *The Growth of Biological Thought: Diversity, Evolution, and Inheritance.* Cambridge, MA: Belknap Press.

Mayr, E. (1988). *Toward a New Philosophy of Biology: Observations of an Evolutionist.* Cambridge, MA: Belknap Press.

Raven, P. H., and G. B. Johnson (1986). *Biology.* St. Louis: Times Mirror/Moseby College.

Rensch, B. (1947). *Neuere Probleme der Abstammungslehre.* Stuttgart: Ferdinand Enke.

Rieppel, O. (1987). "Punctuational Thinking at Odds with Leibniz and Darwin," *N. Jb. Geol. Paleont. Abh.,* Vol. 174, pp. 123–133.

Salvini-Plawen, L. V., and E. Mayr (1977). "On the Evolution of Photoreceptors and Eyes," *Evolutionary Biology,* Vol. 10, pp. 207–263.

Sheppard, P. M. (1975). *Natural Selection and Heredity,* 4th ed. London: Hutchinson.

Simon, H. A. (1962). "The Architecture of Complexity," *Proc. Amer. Phil. Soc.,* Vol. 106, pp. 467–482.

Simpson, G. G. (1944). *Tempo and Mode in Evolution.* New York: Columbia University Press.

Simpson, G. G. (1949). *The Meaning of Evolution: A Study of the History of Life and Its Significance for Man.* New Haven, CT: Yale University Press.

Simpson, G. G. (1953). *The Major Features of Evolution.* New York: Columbia University Press.

Sober, E. (1992). "The Evolution of Altruism: Correlation, Cost, and Benefit," *Biology and Philosophy,* Vol. 7, pp. 177–187.

Stanley, S. M. (1975). "A Theory of Evolution above the Species Level," *Proc. Nat. Acad. Sci.,* Vol. 72:2, pp. 646–650.

Stanley, S. M. (1979). *Macroevolution—Pattern and Process.* San Francisco: W. H. Freeman.

Stebbins, G. L. (1950). *Variation and Evolution in Plants.* New York: Columbia University Press.

Templeton, A. R. (1982). "Adaptation and the Integration of Evolutionary Forces." In *Perspectives in Evolution,* edited by R. Milkman. Sunderland, MA: Sinauer, pp. 15–31.

Tinbergen, N. (1951). *The Study of Instinct.* Oxford: Oxford University Press.

Weiner, N. (1961). *Cybernetics,* Part 2, Cambridge, MA: MIT Press.

Williams, G. C. (1966). *Adaptation and Natural Selection: A Critique of Some Current Evolutionary Thought.* Princeton, NJ: Princeton University Press.

Wilson, E. O. (1992). *The Diversity of Life.* New York: W. W. Norton.

Wooldridge, D. E. (1968). *The Mechanical Man: The Physical Basis of Intelligent Life.* New York: McGraw-Hill.

Wright, S. (1931). "Evolution in Mendelian Populations," *Genetics,* Vol. 16, pp. 97–159.

Wright, S. (1932). "The Roles of Mutation, Inbreeding, Crossbreeding, and Selection in Evolution." In *Proc. 6th Int. Cong. Genetics,* Vol. 1. Ithaca, NY, pp. 356–366.

Wynne-Edwards, V. C. (1962). *Animal Dispersion in Relation to Social Behavior.* London: Oliver and Boyd.

CHAPTER 3

COMPUTER SIMULATION
OF NATURAL EVOLUTION

3.1 EARLY SPECULATIONS
AND SPECIFIC ATTEMPTS

The idea of generating machine learning through simulated evolution has been proposed many times (see Fogel, 1998 for a complete review). Cannon (1932) pictured natural evolution as a learning process, very little different in mechanism or consequence from what an individual undergoes, proceeding by random trial and error. Turing (1950) recognized "an obvious connection between [machine learning] and evolution." Friedman (1956, 1959) speculated that a simulation of mutation and selection would be able to design "thinking machines," and he specifically remarked that chess playing programs could be designed by such a method. Campbell (1960), following earlier work (Campbell, 1956a, b), offered the conjecture that "in all processes leading to expansions of knowledge, a blind-variation-and-selective-survival process is involved." The first attempts to apply evolutionary theory formally to practical engineering problems appeared in the areas of statistical process control, machine learning, and function optimization. Some of these contributions are reviewed here.

3.1.1 Evolutionary Operation

The famous statistician G.E.P. Box and colleagues (Box, 1957, 1960, 1966; Box and Hunter, 1959; Box and Chanmugam, 1962; Box and Draper, 1969) advocated a technique called *evolutionary operation* (EVOP) which was applied to a manufacturing plant as a management process, implemented through the votes of a committee of technical managers. EVOP has been applied successfully to a wide variety of manufacturing processes, although it saw its earliest use in the chemical industry (Hunter, 1958, 1960; Koehler, 1958a, b; Barnett, 1960). Under EVOP, the plant is viewed as an evolving species. The quality of the product advances through variation and selection as determined by the committee (Box and Draper, 1969, pp. 8–9). EVOP is still mentioned occasionally in literature on the optimization of manufacturing processes, but the method was never implemented in autonomous computer simulation.

3.1.2 A Learning Machine

Friedberg (1958; Friedberg et al., 1959) focused attention on gradually improving a machine language computer program, which itself coded for a program, by choosing instructions that were most often associated with a successful result. The number of possible programs for the selected tasks was very large, forbidding an exhaustive search for the best program. The hope was for similar programs to be clustered so as to receive similar judgment. Toward this goal, every program that had a specific instruction in a certain location was termed to be in a *class*. Each contending program belonged to as many different classes as there were locations for instructions. The learning procedure, in evaluating the performance of two nonoverlapping classes of programs, compared instructions occupying the same location. Friedberg devised a scoring technique designed to classify good and bad instructions appropriately.

Scoring was made only with respect to single instructions. This was due mainly to the enormous computer time and storage required to evaluate pairs or higher-ordered combinations of instructions. A program called *Herman* (not an acronym) simulated a computer that processed a set of 64 instructions $(I_0, \ldots, I_{63})$, each comprised of 14 bits. Every combination of 14 bits (six bits for the operation and eight bits for the operand) was an executable instruction, and every sequence of 64 instructions was a performable program. An external program called the *Teacher* caused Herman to be performed successively over many trials and examined Herman's memory to determine whether a desired task had been performed successfully in that trial. Success or failure enabled a third program, the *Learner,* to evaluate the different instructions that appeared in Herman.

Before each new trial, random bits were placed in specific memory locations as inputs, with the others remaining as they were from the previous trial. The program was started at I_0 and terminated upon execution of I_{63}. The Teacher then examined the contents of preselected data locations (outputs) and determined whether or not the bits in these locations satisfied a chosen relationship with the bits placed in the input locations at the beginning of the trial. If the program had not terminated after executing 64 instructions, a failure was declared. Only success or failure was indicated to the Learner.

Because 2^{14} different instructions could occupy a single location I_k, it was impossible to monitor the performance of each instruction given the available computer architecture. Instead, two instructions (initially chosen at random) were "on record" for each location I_k, so that a total of 128 instructions were on record. For each I_k, one of the two instructions on record was "active" and the other was "inactive." In any trial, the program executed by Herman consisted of the 64 active instructions. The Learner altered the program by two methods: (1) it frequently interchanged the two instructions on record for a single location based on their association with successful results (i.e., their class score), and (2) it occasionally made a random change by removing one of the 128 instructions from the record and replacing it with a new 14-bit number. A random change was made after every 64th failure. Random changes were timed with a given number of failures so as to make alterations infrequent when the program was performing optimally.

The results of several experiments indicated that the learning program evolved programs that would output identical or complementary bits to the input, or lower-order bits of the sum of two input bits, but did not achieve success in generating more complicated functions of the input, such as a logical AND. Friedberg et al. (1959) indicated that the length of time required to adapt the program was longer than the expected time required by a completely random search. Changes in systematic order based on success or failure led to a reasonable chance that the process would cycle, creating duplicate programs every 128 failures. Updating Herman at random showed significant improvement but still only approached the efficiency of a completely random search.

To facilitate more efficient learning, Friedberg et al. (1959) suggested a hill-climbing search. The given problem was to be partitioned into parts, with the more difficult parts dealt with first. This approach was termed *Ramsy*. The Teacher ordered the parts in terms of difficulty. If the program failed on a given run, all of the participating (i.e., active) instructions were replaced by new random instructions. If the program succeeded, all of the participants became *bound,* and the next part of the problem was undertaken. Bound instructions were exempted from subsequent replacement unless no participants could be replaced after a failing run because all were bound. Under such conditions, Ramsy was initialized with new instructions and learning was restarted.

Experiments in learning typical logical operators, including AND, EXCLUSIVE-OR, and so forth, were conducted. Ramsy performed better than completely random chance techniques, but its primary drawback was a tendency to become stuck as successive parts of the problem were encountered. This precluded the method from learning solutions to complex programming problems. Friedberg (1958) originally described the research as showing "limited success" but admitted after experiments with Ramsy that "where we should go from here is not entirely clear" (Friedberg et al., 1959).

Although Friedberg et al. (1959) did not explicitly claim to be simulating natural evolution, that is the generally accepted interpretation (Jackson, 1974). Dunham and North, Friedberg's (1959) co-authors, observed that simulated evolution is "more powerful than commonly recognized" and cited success in evolving bit manipulators and in laying out the wiring of a 1400-terminal black box (Dunham et al., 1963), which served as the inertial guidance system for a missile. These successes were not given much attention by the artificial intelligence community.

3.2   ARTIFICIAL LIFE

Perhaps the earliest record of any effort in evolutionary computation was due to Barricelli (1954), who worked on the high-speed computer at the Institute for Advanced Study in Princeton, New Jersey, in 1953. The experiments conducted were essentially similar to those popularized over 30 years later in the field of "artificial life." Barricelli's original research was published in Italian but was subsequently republished as Barricelli (1957) in English and was followed up in Barricelli (1962, 1963). In particular, Barricelli's experiments involved a simulation of numeric elements in a grid. These numbers propagated by means of local rules, such as positive numbers shifted to the right in the grid while negative numbers shifted left. When collisions occurred between two or more numbers entering the same cell, various rules were employed to alter the numeric elements. Barricelli was particularly interested in what "emergent patterns" would be observed from these simple simulations. Figure 3-1 shows a series of generations under the basic reproduction and mutation rules of Barricelli (1957). Barricelli showed that synergistic patterns could be generated that worked together to copy themselves into subsequent generations. For more detail, see Fogel (1998, pp. 163–164).

Conrad and Pattee (1970) also studied general properties of evolution sys-

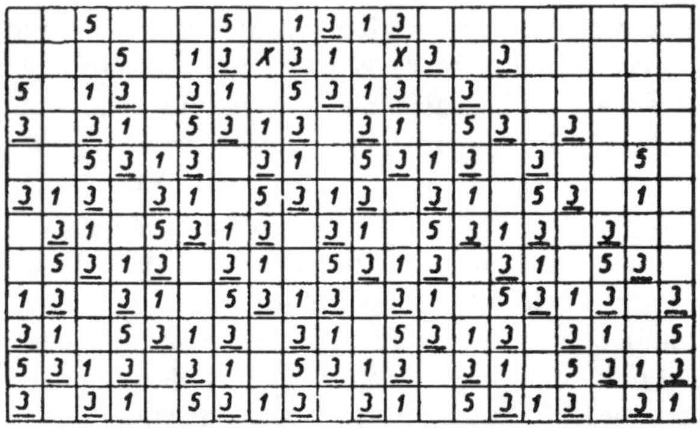

Figure 3-1 A series of generations under the basic reproduction and mutation rules of Barricelli (1957). The first generation is represented as the top row of cells. Subsequent generations are shown as successive descending rows of cells. The numeric elements propagate according to the rules described in the text. By the fourth generation, the pattern $(5, -3, 1, 0, -3, 1)$ appears and persists in all subsequent generations. Note that a number with an underscore has a negative value. The figure is adapted from Barricelli (1957).

tems but from a different perspective that involved the simulation of a hierarchic ecosystem. A population of cell-like individual organisms placed in an array of environmental cells was subjected to a strict materials conservation law that induced competition for survival. The organisms were capable of mutual cooperation, as well as executing biological strategies that included genetic recombination and the modification of the expression of their genome. No fitness criteria were introduced explicitly as part of the program. Instead, the simulation was viewed as an ecosystem in which genetic, individual, and populational interactions would occur and behavior patterns would emerge (see Fogel, 1998, pp. 403–405 for more details).

Conrad (1981) and Rizki and Conrad (1985) conducted further studies of population dynamics in hierarchic ecosystems. In the latter effort, simulated organisms were given 15 phenotypic traits, including temperature optimum, temperature tolerance, rate of energy intake, adult period, and mutation rate. A pleiotropic/polygenic coding scheme was utilized (for details, see Rizki and Conrad, 1985). It was concluded that "populations cultured in constant environments usually developed adaptability more rapidly than those cultured in variable environments. . . . However, the situation was reversed when the populations are cultured in the variable environment long enough for them to develop suitable adaptations" (Rizki and Conrad, 1985). Populations were also seen to vary as a result of sensitivity to initial conditions. These properties were attributed to "founder effects" (see Mayr, 1963, for discussion). Additional experiments were reported in Conrad and Rizki (1989).

Another line of investigation into evolutionary dynamics was emphasized by Ray (1991a, b), who conducted experiments within the framework of an open-ended ecology. Ray (1991a) focused on developing evolutionary simulations that would permit study of features that are fundamental to all evolutionary systems, not just organic systems. Within that framework, he devised a simulation (*Tierra*) that competes computer programs written in assembly code. The initial program consisted of three parts: (1) a self-exam routine that determined the total length of the program, (2) a reproduction loop, and (3) a copy procedure that would be called by the reproduction loop for as many iterations as there were instructions in the program to replicate. The memory, the central processing unit (CPU), and the computer's operating system formed the physical environment. The CPU was allocated to programs in the environment such that shorter programs would replicate faster than longer programs. Programs that generated errors were removed (actually, simply deallocated) from the environment based on a function of the number of failed instructions attempted. Failures occurred because a mutation operation was employed that randomly altered instructions both in a general background mode, where every bit in every program in the simulation was given a small probability of mutation, and in a copy mode, where additional alterations could be imposed during self-replication.

An 80-instruction handcrafted program that would replicate itself was used to seed the simulation. Over successive generations, the evolutionary dynamics invented a number of interesting programs with different functionalities. Ray (1991a, b) described one class of program as being *parasitic;* that is, it used the copy procedures of other programs to replicate its own code, thereby reducing its own program length. (It only had a self-exam and reproduction loop.) Because these shorter programs required less CPU time to complete replication, they were able to outcompete their longer *host* programs. Subsequently, *hyperparasites* were created, which allowed the parasites to access their copy procedure but then usurped control of the reproduction loop and used their own reproduction loop to copy themselves instead of the parasites. Other emergent behaviors are described in Ray (1991a, b).

Tierra demonstrates that complex behaviors can arise from evolutionary dynamics operating on relatively simple competing programs. Assessing the functionality of any particular program is tedious because it involves a line-by-line examination of the assembler code of the program and requires an assessment of how it interacted with the other programs or residual code left over from programs that had perished. Nevertheless, Ray (personal communication) is working toward implementing Tierra over the Internet, with the hope of understanding important features associated with the evolution of the division of functionality in organisms.

3.3   EVOLUTIONARY PROGRAMMING

L. Fogel proposed using simulated evolution on a population of contending algorithms in order to develop artificial intelligence, and he explored this possibility of *evolutionary programming* in a series of studies (Fogel, 1962a, b, c, 1963, 1964; Fogel and Walsh, 1964; Fogel et al., 1965a, b, 1966; and others). Intelligent behavior was viewed as requiring the composite ability: (1) to predict one's environment, coupled with (2) a translation of the predictions into a suitable response in light of the given goal. For the sake of generality, the environment was described as a sequence of symbols taken from a finite alphabet. The evolutionary problem was defined as evolving an algorithm that would operate on the sequence of symbols thus far observed in such a manner as to produce an output symbol that would likely maximize the algorithm's performance in light of the next symbol to appear in the environment and a well-defined payoff function. Finite-state machines provided a useful representation for the required behavior.

A finite-state machine is a transducer that can be stimulated by a finite alphabet of input symbols, can respond in a finite alphabet of output symbols, and possesses a finite number of different internal states. The corresponding input–output symbol pairs and next-state transitions for each input symbol, taken over every state, specify the behavior of any finite-state machine, given any starting state. For example, a three-state machine is shown in Figure 3-2.

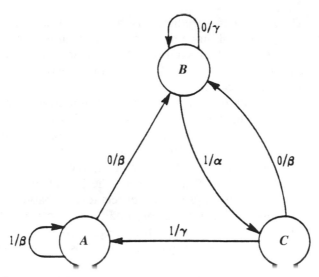

Figure 3-2 A three–state finite-state machine. Input symbols are shown to the left of the virgule. Output symbols are to the right of the virgule. Unless otherwise specified, the machine is presumed to start in state "A" (after Fogel et al., 1966, p. 12).

The alphabet of input symbols comprises $\{0, 1\}$, whereas the alphabet of output symbols comprises $\{\alpha, \beta, \gamma\}$. (Input symbols are shown to the left of the virgule, and output symbols are shown to the right.) The finite-state machine transforms a sequence of input symbols into a sequence of output symbols. Table 3-1 indicates the machine's response to a given string of symbols, presuming that the machine is found in the state C. It is presumed that the machine acts when each input symbol is perceived and the output takes place before the next input symbol arrives.

Fogel proposed evolutionary programming to operate on finite-state machines in the following manner. A population of finite-state machines is exposed to the environment—that is, the sequence of symbols that have been observed up to the current time. For each parent machine, as each input symbol is offered to the machine, each output symbol is compared to the next

Table 3-1 The Response of the Finite-State Machine Shown in Figure 3-2 to a String of Symbols. In This Example, the Machine Starts in State C.

Present State	C	B	C	A	A	B
Input Symbol	0	1	1	1	0	1
Next State	B	C	A	A	B	C
Output Symbol	β	α	γ	β	β	α

input symbol. The worth of this prediction is then measured with respect to the given payoff function (e.g., all-none, absolute error, squared error, or any other expression of the meaning of the symbols). After the last prediction is made, a function of the payoff for each symbol (e.g., average payoff per symbol) indicates the fitness of the machine.

Offspring machines are created by randomly mutating each parent machine. Each parent produces a single offspring (simply for convenience). Five possible modes of random mutation result naturally from the description of the machine: change an output symbol, change a state transition, add a state, delete a state, or change the initial state. The deletion of a state and the changing of the initial state are allowed only when the parent machine has more than one state. Mutations are chosen with respect to a probability distribution, which is typically uniform. The number of mutations per offspring is also chosen with respect to a probability distribution (e.g., Poisson) or may be fixed *a priori*. These offspring are then evaluated over the existing environment in the same manner as their parents. Other mutations, such as majority logic mating operating on three or more machines, were proposed in Fogel et al. (1966) but not implemented.

Those machines that provide the greatest payoff are retained and become parents of the next generation. Typically, half of the total machines are saved so that the parent population remains at a constant size. This process is iterated until an actual prediction of the next symbol (as yet unexperienced) in the environment is required. The best machine generates this prediction, the new symbol is added to the experienced environment, and the process is repeated. Fogel (1964; Fogel et al., 1966) used "nonregressive" evolution. A machine had to rank in the top half of the population in order to be retained. Saving lesser adapted machines was discussed as a possibility (Fogel et al., 1966, p. 21) but not incorporated.

Such a procedure has an inherent versatility. The payoff function can be arbitrarily complex and can possess temporal components; there is no requirement for the classical squared error criterion or any other smooth function. Moreover, it is not required that the predictions be made with a one-step look ahead. Forecasting can be accomplished at an arbitrary length of time into the future. Multivariate environments can be handled, and the environmental process need not be stationary because the simulated evolution will adapt to changes in the transition statistics.

Fogel et al. (1964, 1966) described a series of experiments in which successively more difficult prediction tasks were presented to the evolutionary program. These included simple two-symbol cyclic sequences, more complex eight-symbol cyclic sequences degraded by random noise, sequences of symbols generated by other finite-state machines, nonstationary sequences, and sequences taken from Flood (1962).

For example, a nonstationary sequence of symbols was generated by classifying each of the increasing integers as being prime (symbol 1) or non-prime (symbol 0). Thus, the environment consisted of the sequence 01101010001 . . . , where each symbol depicts the primeness of the positive integers 1, 2, 3, 4, 5, 6, 7, 8, 9, 10, 11, . . . , respectively. Testing for primeness is straightforward, but predicting whether or not the next integer will be prime based on the sequence of observed primes and nonprimes appeared nontrivial. The payoff function for prediction was all-or-none—that is, one point for each correct prediction, zero points for each error, modified by subtracting 0.01 multiplied by the number of states of the machine. This penalty for complexity was provided to maintain parsimonious machines in light of the limited memory of the available computer (IBM 704).

Figure 3-3 shows the cumulative percentage of correct predictions in the first 200 symbols. After the initial fluctuation (due to the small sample size), the prediction score increased to 78 percent at the 115th symbol and then essentially remained constant until the 200th prediction. At this point, the best machine possessed four states. At the 201st prediction, the best machine possessed three states, and at the 202nd prediction, the best machine possessed only one state, with both output symbols being 0. After 719 symbols, the process was halted, with the cumulative percentage of correct predictions reaching 81.9 percent. The asymptotic worth of this machine would be 100 percent correct because the prime numbers become increasingly infrequent and the machine continued to predict "nonprime."

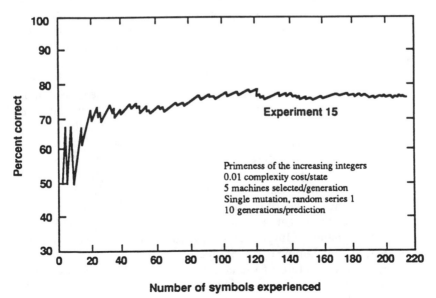

Figure 3-3 The cumulative percentage of correct predictions in the first 200 prime numbers (from Fogel et al., 1966, p. 37).

The goal was then changed to offer a greater payoff for predicting a rare event. Correctly predicting a prime was worth one plus the number of nonprimes that preceded it. Similarly, correctly predicting a nonprime was worth one plus the number of primes that preceded that nonprime. During the first 150 symbols there were 30 correct predictions of primes, 37 incorrect predictions (false alarms), and 5 missed primes. From the 151st symbol to the 547th there were 65 predictions of primes and 67 false alarms. That is, of the first 35 primes, 5 were missed—of the next 65 primes, none was missed. Fogel et al. (1966) indicated that the evolutionary algorithm quickly learned to recognize numbers that are divisible by two or three as being nonprime. Some recognition that numbers divisible by five are nonprime was also evidenced. Fogel (1968) later remarked that the evolutionary program had successfully discovered "cyclic properties within the environment . . . in essence, the equivalent of first recognizing that numbers divisible by two are not prime, etc. In other words, the program was synthesizing a definition of primeness without prior knowledge of the nature of primeness *or* an ability to divide. Clearly, the evolutionary program would never be a perfect predictor, but that is to be expected." No finite-state machine can represent the algorithm for generating prime numbers.

Fogel (1962b, 1966) considered the problem of designing conscious automata. Consciousness was viewed as self-referential modeling. To be conscious, "the entity must include within itself an ability to develop various models of itself, to examine these models upon request, to select a suitable model in the light of the inquiry and to display that model to the observer upon request" (Fogel, 1966). Such modeling was made through the use of evolving finite-state models of evolving finite-state predictors coupled with the environment.

Fogel and Burgin (1969) used evolutionary programming for gaming (coevolution). A series of two-person zero-sum gaming experiments was conducted. Evolutionary programming was able to discover the globally best strategy in simple games involving a small number of possible plays (e.g., four) and outperform human subjects in more complicated circumstances. Extensions addressed nonzero-sum games such as pursuit-evasion. Evolutionary programming was utilized to guide an interceptor towards a moving target in three dimensions. Because the target was generally allowed greater maneuverability, the success of the interceptor depended on an ability to predict the position of the target at some future time without *a priori* knowledge of the target's dynamics. Burgin (1969) provided further description of the two-person zero-sum gaming experiments.

The time-varying nature of the adaptive topography in the cited experiments (1962–1970) helps prevent evolutionary stagnation. After each prediction, the newly observed symbol is added to the experienced sequence. The fitness of the evolving machines reflects the payoff from the last prediction. The adaptive topography changes. Machines that perform poorly on the last prediction are mapped onto a higher position on the error surface, whereas machines that offer good predictions have a corresponding lower error score. Because the adaptive topography is in continual flux, the problems associated with evolutionary stagnation on a static topography are reduced considerably.

The representation becomes increasingly complex as the evolving finite-state machines discover regularities in the experienced sequence. A larger number of states provides for a greater degree of expressible behavior (i.e., possible sequences). This complexity has other advantages as well. Neutral mutations (e.g., addition of a state that is unexpressed when the machine is driven by the observed environment) may serve as an informational buffer, reducing the chance that random mutation will change expressed states. Furthermore, the change in, say, a single output symbol is likely to have less phenotypic (behavioral) impact on a 20-state machine than on a two-state machine, assuming that the change is expressed. Increasing the number of states has the effect of reducing the step size of the search, allowing for fine-tuning the evolving structure to the bottom of a minimal error trough.

Additional evolutionary programming experiments in sequence prediction were performed by Lutter and Huntsinger (1969). Evolutionary modeling was able to rapidly relearn environments that had been previously experienced, even after radical environmental change. The sequence of finite-state machines had built in a memory of the previous sequences. Burgin (1974) offered a comparison of evolutionary system identification to quasilinearization and indicated that evolutionary programming was well suited for modeling systems described by nonlinear differential equations in light of nontraditional payoff functions. Dearholt (1976) successfully used evolutionary classification techniques on samples of electrocardiograms depicting common heart diseases.

Initial reviews of the evolutionary programming method were favorable. Feigenbaum (1963) described the process as building "complex Markovian models." Kleyn (cited in Fogel et al., 1965a) described the process as an "excellent and novel approach to a learning or self organizing system." But Solomonoff (1966) remarked that "the method of Fogel et al. should not be regarded in its present state as being particularly good for working any but the simplest problems." Solomonoff described evolutionary programming mainly as a simple hill-climbing process using a single parent machine to evolve a single offspring machine. Little emphasis was placed on the parallel use of multiple machines as was actually performed. This incorrect description persisted in subsequent reviews (e.g., Holland, 1975, p. 163; Kieras, 1976; Rada, 1981; Lenat, 1983).

Solomonoff (1966) had a number of other criticisms. He commented that "a relatively minor [difficulty] is the penalty for complexity of machine that is incorporated into the 'score' that the machine makes on a prediction run."

Fogel et al. (1966) generally set such penalty terms to be proportional to the number of states. Solomonoff (1966) suggested that the appropriate penalty term "is by no means arbitrary, but can be computed—in some cases with exactness. Use of any other penalty function will result, on the average, in poorer predictions."

Such criticism is well directed, but unfortunately it is often difficult to compute the exact penalty term. Mallows (1973), Akaike (1974), Rissanen (1978), Barron (1984), and others have suggested procedures for assessing the tradeoff between the goodness of fit of a model and the number of its associated degrees of freedom. Assumptions such as the environmental noise being stationary and Gaussian are typically required. But Fogel et al. (1966) were specifically interested in nonstationary, non-Gaussian environments and the appropriate choice for a penalty for overfitting the observed data remains debatable (Fogel, 1991a). Furthermore, the penalty terms in both Akaike's information criterion (Akaike, 1974) and the minimum description length principle (Rissanen, 1978) are essentially proportional to the degrees of freedom of the model, as was the penalty term employed by Fogel et al. (1966).

Solomonoff was concerned that the simulated evolution did not contain overt mechanisms corresponding to observed genetic processes such as crossover and inversion. He commented that such mechanisms tend to "push synergistic genes together on the chromosome"; he defined synergistic as "two things are 'synergistic' if their occurrence together produces more beneficial effect than the sum of the benefits they give when they occur separately." But such a view requires assigning a fitness to individual genes. As discussed in Chapter 2, this is an inappropriate perspective.

Solomonoff described research in simulated evolution to be "very limited in both quantity and quality." He further speculated that "while it is possible to develop machines to solve simple problems by mutating a few states at a time, this process is far too slow to obtain solutions to complex problems." No firm description of what was meant by slow or complex was offered. He did, however, encourage further research in the area.

In order to make a preliminary assessment of the lack of complexity attributed by Solomonoff (1966), the size of the available state space of possible solutions can be computed. Atmar (1976) calculated that the number of possible configurations for a finite-state machine given n states, a input symbols, and b output symbols is given by

$$N = (n^a b^a)^n.$$

Table 3-2 indicates the number of possible machines for $n \leq 10$, and input/output alphabets of 1, 2, 3, 4, and 8 symbols. Fogel et al. (1966) conducted experiments in which the environment was taken to be a binary sequence and the maximum number of states for any machine was set at 25. Therefore, the size of the state space was $(25^2 2^2)^{25} = (5^{100})(2^{50})$. The state space becomes immensely large, even when restricted to relatively small machines.

Table 3-2 The Number of Possible Finite-State Machines Given n States and $\alpha = \beta$ Input–Output Symbols.

	$\alpha/\beta = 1$	$\alpha/\beta = 2$	$\alpha/\beta = 3$	$\alpha/\beta = 4$	$\alpha/\beta = 8$
$n = 1$	1	4	27	256	1.7×10^7
$n = 2$	4	256	$46,656$	1.7×10^7	1.8×10^{19}
$n = 3$	27	$46,656$	3.9×10^9	9.0×10^{12}	1.3×10^{33}
$n = 4$	256	1.7×10^7	9.0×10^{12}	1.9×10^{19}	1.5×10^{48}
$n = 5$	$3,125$	1.0×10^{10}	4.4×10^{17}	1.1×10^{26}	1.2×10^{64}
$n = 6$	$46,656$	9.0×10^{12}	4.0×10^{22}	1.3×10^{33}	5.0×10^{80}
$n = 7$	8.2×10^5	1.1×10^{16}	5.8×10^{27}	3.3×10^{40}	7.3×10^{98}
$n = 8$	1.7×10^7	1.8×10^{19}	1.3×10^{33}	1.5×10^{48}	3.9×10^{115}
$n = 9$	3.9×10^9	3.9×10^{22}	4.4×10^{38}	1.1×10^{56}	1.2×10^{126}
$n = 10$	1.0×10^{10}	1.1×10^{26}	2.1×10^{44}	1.2×10^{64}	1.8×10^{152}

Subsequent reviews were mixed. Brody and Lindgren (1968) remarked that "although such simulation of evolution by computer . . . has thus far dealt with only relatively simple problems, the method appears to offer an extremely powerful tool." Lutter and Huntsinger (1969) felt that evolutionary programming was especially suited to interpret continuous data from a stochastic process and hoped "to encourage the use of these techniques, especially in control applications." But Michie (1970) wrote: "The idea of getting a computer to write its own program has appeared and disappeared several times in the past two decades. Early attempts, inspired by the example of biological evolution, were based on generating program symbols randomly, conserving the more successful sequences. Such an approach is now considered naive." Similar positive comments were offered by Atmar (1976), and more critical general comments were offered by Chandrasekaran and Recher (1974), Jackson (1974), and Rada (1981).

Lenat (1983) put forth a strong criticism:

The early (1958–1970) researchers in automatic programming were confident that they could succeed by having programs randomly mutate into desired new ones. This hypothesis was simple, elegant, aesthetic, and incorrect. The amount of time necessary to synthesize or modify a program was seen to increase exponentially with its length. Switching to a higher-level language . . . merely chipped away somewhat at the exponent, without muffling the combinatorial nature of the process. All the attempts to get programs to "evolve" failed miserably, casualties of the combinatorial explosion.

But this claim that the amount of time necessary to evolve a solution increases exponentially with its length is unsubstantiated in the literature. Investigations by Ambati et al. (1991) and Fogel (1993a) indicate to the contrary.

Perhaps the most pointed criticism of Fogel et al. (1966) was offered by Lindsay (1968). A random search, Lindsay observed, "is . . . the most inefficient method of problem solving," and as evidence, he cited the failure of Friedberg (1958). He proceeded to claim incorrectly that the evolutionary search of Fogel et al. (1966) was no better than completely random search, and he concluded by proclaiming the work of Fogel et al. (1966) to be "fustian" that "may unfortunately alienate many psychologists from the important work being done in artificial intelligence." Readers interested in a more careful analysis of the criticisms of early evolutionary programming are referred to Fogel (1991a, 1992).

Fogel (1995) reported that only limited work in evolutionary programming was performed during the 1970s and 1980s (e.g., Atmar, 1976; Takeuchi, 1980). It has been recently discovered, however, that a series of master's theses were explored at New Mexico State University during this time frame, including Root (1970), Cornett (1972), Lyle (1972), Holmes (1973), Trellue (1973), Montez (1974), Vincent (1976), and Williams (1977). These efforts examined primarily handwritten character recognition and/or the induction of languages using Moore machines. Thus, in contrast to the remarks in Fogel (1995), evolutionary programming received considerable attention during this time frame.

More recently, the technique has been applied to diverse combinatorial optimization problems. Rather than use finite-state machines, representations are chosen based on the problem at hand, and variation operations are constructed that maintain a strong behavioral linkage between each parent and its progeny. The procedure has been applied, for example, to path planning problems (see early efforts in Fogel, 1988, 1993a, b; McDonnell and Page, 1990; Fogel and Fogel, 1990; McDonnell et al., 1992; Page et al., 1992), the training and design of neural networks (e.g., Fogel et al., 1990; McDonnell and Waagen, 1994; Angeline et al., 1994; Yao and Liu, 1997; Fogel et al., 1998; Sebald and Chellapilla, 1998), automatic control (Sebald and Schlenzig, 1994), gaming (Fogel, 1991b, 1993c; Porto and Fogel, 1997), and general function optimization (Fogel and Atmar, 1990; Fogel and Stayton, 1994). Considerable efforts have been made in the application to problems in power systems (Lai and Ma, 1996; Lai et al., 1996; Yang et al., 1996; Wong and Yuryevich, 1998; Lai, 1998). The potential benefits of parallelizing evolutionary programming was examined initially by Duncan (1993) and Nelson (1994). The most recent efforts can be found in McDonnell et al. (1995), Fogel et al. (1996), Angeline et al. (1997), and Porto et al. (1998) (also see Fogel and Atmar, 1992, 1993 and Sebald and Fogel, 1994).

3.4 EVOLUTION STRATEGIES

Another approach to simulating evolution was adopted independently by Rechenberg (1965) and Schwefel (1965), collaborating in Germany along with another graduate student, Peter Bienert, in 1964 at the Technical University of Berlin. They initially addressed optimization problems in fluid mechanics as generated by hardware devices, but they turned toward general function optimization algorithms. The simplest version of their "evolution strategy" can be implemented as follows:

1. The problem is defined as finding the real-valued n-dimensional vector r that is associated with the extremum of a functional $F(x): R^n \to R$. Without loss of generality, let the procedure be implemented as a minimization process.

2. An initial population of parent vectors, $x_i, i = 1, \ldots, P$, is selected at random from a feasible range in each dimension. The distribution of initial trials is typically uniform.

3. An offspring vector, x_i' is created from each parent $x_i, i = 1, \ldots, P$, by adding a Gaussian random variable with zero mean and preselected standard deviation to each component of x.

4. Selection then determines which of these vectors to maintain by ranking the errors $F(x_i)$ and $F(x_i'), i = 1, \ldots, P$. The P vectors that possess the least error become the new parents for the next generation.

5. The process of generating new trials and selecting those with least error continues until a sufficient solution is reached or the available computation is exhausted.

In this model, the components of a trial solution are viewed as behavioral traits of an individual, not as genes along a chromosome (cf. Bremermann, 1962). A genetic source for these phenotypic traits is presumed, but the nature of the linkage is not detailed. It is assumed that whatever genetic transformations occur, the resulting change in each behavior will follow a Gaussian distribution with zero mean difference and some standard deviation. Specific genetic alterations can affect many phenotypic characteristics due to pleiotropy and polygeny (Figure 2-2). It is therefore appropriate to simultaneously vary all components of a parent in the creation of a new offspring.

The original efforts in evolution strategies (Rechenberg, 1965; Schwefel,

1965) examined the preceding algorithm but focused on a single parent–single offspring search. This was termed a $(1 + 1)$–ES in that a single offspring is created from a single parent and both are placed in competition for survival, with selection eliminating the poorer solution. This approach had two main drawbacks when viewed as a practical optimization algorithm: (1) the constant standard deviation (average step size) in each dimension made the procedure slow to converge on optimal solutions, and (2) the brittle nature of a point-to-point search made the procedure susceptible to stagnation at local minima (although the procedure can be shown to asymptotically converge to the global optimum vector x; see Solis and Wets, 1981).

Rechenberg (1973) defined the expected convergence rate of the algorithm as the ratio of the average distance covered toward the optimum and the number of trials required to achieve this improvement. For a quadratic function:

$$F(x) = \Sigma \, x_i^2, \tag{3.1}$$

where x is an n-dimensional vector of reals and x_i denotes the ith component of x, Rechenberg (1973) demonstrated that the optimum expected convergence rate is given when $\sigma \approx 1.224r/n$, where σ is the standard deviation of the zero mean Gaussian perturbation, r denotes the current Euclidean distance from the optimum, and there are n dimensions. Thus, for this simple function the optimum convergence rate is obtained when the average step size is proportional to the square root of the error function and inversely proportional to the number of variables. Additional analyses have been conducted on other functions, and the results have yielded similar forms for setting the standard deviation (Bäck et al., 1991) (see Chapter 4).

Schwefel (1981) developed the use of multiple parents and offspring in evolution strategies, following earlier work of Rechenberg (1973) that used multiple parents but only a single offspring. More recently, two main approaches have been explored, denoted by $(\mu + \lambda)$–ES and (μ, λ)–ES. In the former, μ parents are used to create λ offspring and all solutions compete for survival, with the best μ being selected as parents of the next generation. In the latter, only the λ offspring compete for survival and the μ parents are completely replaced each generation. That is, the life span of every solution is limited to a single generation. Increasing the population size increases the rate of optimization over a fixed number of generations.

To provide a very simple example, suppose it is desired to find the minimum of the function in Eq. 3.1 for $n = 3$. Let the original population consist of 30 parents, with each component initialized in accordance with a uniform distribution over $[-5.12, 5.12]$ (after De Jong, 1975). Let one offspring be created from each parent by adding a Gaussian random variable with mean zero and variance equal to the error score of the parent divided by the square of the number of dimensions ($3^2 = 9$) to each component. Let selection simply retain the best 30 vectors in the population of parents and offspring. Figure 3-4 indicates the rate of optimization of the best vector in the population as a

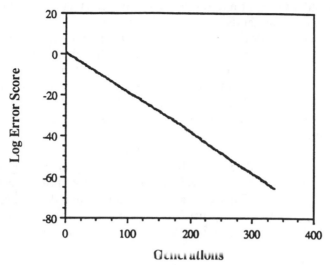

Figure 3-4 Using an evolution strategy to find the minimum of a simple three–dimensional quadratic surface. A population of 30 real-valued vectors is maintained at each generation. The curve depicts the rate of optimization of the best vector in the population at each generation. The error score is on a logarithmic scale.

function of the number of generations. The process rapidly converges close to the unique global optimum.

Rather than use a heuristic schedule for reducing the step size over time, Rechenberg introduced the idea of making the distribution of new trials from each parent an additional self-adaptive parameter.[1] In this procedure (detailed in Schwefel, 1981), each solution vector comprises not only the trial vector x of n dimensions, but a perturbation vector σ that provides instructions on how to mutate x and is itself subject to mutation. For example, if x is the current position vector and σ is a vector of variances corresponding to each dimension of x, then a new solution vector (x', σ') could be created as:

$$\sigma_i' = \sigma_i \cdot \exp(\tau' \cdot N(0,1) + \tau \cdot N_i(0,1))$$

$$x_i' = x_i + N(0, \sigma_i')$$

where $i = 1, \ldots, n$, and $N(0,1)$ represents a single standard Gaussian random variable, $N_i(0,1)$ represents the ith independent identically distributed (i.i.d.) standard Gaussian, and τ and τ' are operator-set parameters affecting global and individual step sizes (Bäck and Schwefel, 1993). In this manner, the evolution strategy can adapt to the width of the error surface on-line and more appropriately distribute trials. This method was extended again (see Schwefel, 1981) to incorporate correlated mutations so that the distribution of new

[1]Rechenberg indicated in personal communication that he first wrote down the idea for self-adaptive parameters in 1967.

trials could adapt to contours on the error surface (Figure 3-5). Greater consideration to these methods is given in Chapter 4.

Additional extensions included methods for recombining individual solutions in the creation of new offspring. There are many proposed procedures, such as selecting individual components from either of two parents at random, averaging individual components from two or more parents with a given weighting, and so forth (Bäck et al., 1991). Also possible is giving each individual a certain maximum lifetime of generations, so as to prevent stagnation (Schwefel and Rudolph, 1995).

Strong similarities exist between evolution strategies and evolutionary programming as applied to real-valued continuous optimization problems. In many cases, the procedures are virtually equivalent, although they were developed independently. The extension to self-adapting independent variances in evolutionary programming was offered in Fogel et al. (1991), with procedures for optimizing the covariance matrix used in generating new trials presented in Fogel et al. (1992). These methods differed from those offered in Schwefel (1981) in that Gaussian perturbations are applied to the self-adaptive parameters instead of lognormal perturbations. Initial comparisons (Bäck et al., 1993; Saravanan and Fogel, 1994; Saravanan et al., 1995) indicate that the procedures in Schwefel (1981) appear to be more robust than those in Fogel et al. (1991) on static functions but are less robust on noisy functions (Angeline, 1996). Theoretical and empirical comparison between these mechanisms is an open area of research.

Current efforts in evolution strategies are aimed at developing mathematical foundations for the procedures (Bäck et al., 1993; Rudolph, 1994, 1997;

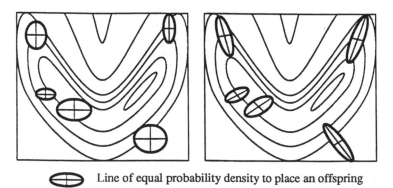

⊕ Line of equal probability density to place an offspring

Figure 3-5 By implementing correlated mutations across various dimensions, the evolutionary search can adapt to arbitrarily oriented contours on the response surface. The first illustration shows the probability density contours for uncorrelated mutations, while the second indicates the potential for moving more rapidly toward minima when using correlated mutations (from Bäck et al., 1991).

Beyer, 1997), and investigating their computational complexity theoretically and empirically (Born et al., 1992; Ostermeier, 1992). Efforts have been made to incorporate evolution strategies in engineering applications (Cai and Thierauf, 1996; Gross et al., 1996; Moosburger et al., 1997; Richter and Hofmann, 1997; Franzen et al., 1998), in designing neural structures (Lohmann, 1992; Ogawa et al., 1997), and in modeling and system identification (Kuhn and Visser, 1992; Fathi-Torbaghan and Hildebrand, 1997). Hoffmeister (1991) has examined parallelizing evolution strategies on distributed processing hardware. Recent investigations in evolution strategies can be found in Voigt et al. (1996) and Eiben et al. (1998) (also see Männer and Manderick, 1992 and Davidor et al., 1994).

3.5 GENETIC ALGORITHMS

An alternative method of simulating evolution, later known as genetic algorithms, arose at least three times independently over the course of approximately a decade. This method was first offered in Fraser (1957a, b, 1960, 1962, 1967, 1968), in Bremermann (1962; Bremermann et al., 1966), and in publications of John Holland's students at the University of Michigan (e.g., Bagley, 1967; Rosenberg, 1967; see also Holland, 1969, 1973). Essentially, the technique (1) incorporated a population of individuals encoded as "chromosomes," mainly in a binary representation (i.e., 1s and 0s), (2) propagated copies of these individuals based on external fitness criteria, and (3) generated new individuals for each next generation by mutating bits and recombining elements from different members of the population.

Genetic algorithms are typically implemented in canonical form as follows:

1. The problem to be addressed is defined and captured in an objective function that indicates the fitness of any potential solution.

2. A population of candidate solutions is initialized subject to certain constraints. Typically, each trial is coded as a vector x, termed a *chromosome*, with elements being described as *genes* and varying values at specific positions called *alleles*. Holland (1975, pp. 70–72) speculated that there would be an advantage to representing individuals by binary strings. For example, if it were desired to find the scalar value x that maximizes:

$$F(x) = -x^2,$$

then a finite range of values for x would be selected and the minimum possible value in the range would be represented by the string $[0 . . . 0]$, with the maximum value being represented by the string $[1 . . . 1]$. The desired degree of precision would indicate the appropriate length of the binary coding.

3. Each chromosome, x_i, $i = 1, \ldots, P$, in the population is decoded into a form appropriate for evaluation and is then assigned a fitness score, $\mu(x_i)$ according to the objective.

4. Each chromosome is assigned a probability of reproduction, p_i, $i = 1, \ldots, P$, so that its likelihood of being selected is proportional to its fitness relative to the other chromosomes in the population. If the fitness of each chromosome is a strictly positive number to be maximized, this is traditionally accomplished using *roulette wheel selection* (Figure 3-6), although other techniques are available.

5. According to the assigned probabilities of reproduction, p_i, $i = 1, \ldots, P$, a new population of chromosomes is generated by probabilistically selecting strings from the current population. The selected chromosomes generate offspring via the use of specific genetic operators, such as crossover and bit mutation.

6. The process is halted if a suitable solution has been found or if the available computing time has expired. Otherwise the process proceeds to step (3), where the new chromosomes are scored and the procedure iterates.

Many forms of crossover were offered by Fraser, Bremermann, and Holland. Holland's was the most straightforward: crossover is applied to two chromosomes (parents) and creates two new chromosomes (offspring) by selecting a random position along the coding and splicing the section that appears before the selected position in the first string with the section that appears after the selected position in the second string, and vice versa (Figure 3-7). Other more sophisticated crossover operators were introduced in Fraser

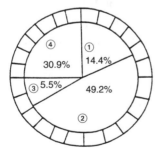

No.	String	Fitness	% of Total
1	01101	169	14.4
2	11000	576	49.2
3	01000	64	5.5
4	10011	361	30.9
Total		1170	100.0

Figure 3-6 Roulette wheel selection in genetic algorithms selects chromosomes for reproduction in proportion to their relative fitness. Essentially, the procedure allocates space on a roulette wheel to each chromosome such that the probability of it being selected varies directly with its fitness (from Goldberg, 1989).

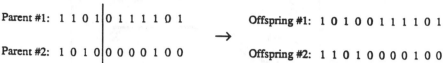

Figure 3-7 The one-point crossover operator is applied to two parents. A crossover point is selected, and the two strings that follow that point are exchanged. Other crossover operators have been offered which rely on multiple crossover points.

(1957a), where each position along the chromosome could have a chance of being a splicing point, and in Bremermann et al. (1965, 1966; Bremermann, 1967, 1968, 1973) where multiparent recombination operators such as majority logic and uniform sampling from various parents were explored.

To consider a simple example, suppose the task is to find a vector of 100 bits {0, 1} such that the sum of all the bits in the vector is maximized. The objective function could be written as

$$\mu(x) = \Sigma \, x_i,$$

where x is a vector of 100 symbols from {0, 1}. Any such vector x could be scored with respect to $\mu(x)$ and would receive a fitness ranging from zero to 100. Let an initial population of 100 parents be selected completely at random and subjected to roulette wheel selection in light of $\mu(x)$, with the probabilities of one-point crossover and bit mutation being 0.8 and 0.01, respectively. Figure 3-8 shows the rate of improvement of the best vector in the population and the average of all parents, at each generation (one complete iteration of steps 3–6) under such conditions. The process rapidly converges on vectors of all 1s.

A number of issues must be addressed when using a genetic algorithm. For example, the suggestion for using binary codings has received considerable criticism (Antonisse, 1989; Vignaux and Michalewicz, 1991; Michalewicz, 1992; Fogel and Ghozeil, 1997). To understand the historical motivation for using bit strings, the notion of schemata must be introduced. Consider a string of symbols from an alphabet A. Suppose that some of the components of the string are held fixed while others are free to vary. Define a wild card symbol, $\# \notin A$, that matches any symbol from A. A string with fixed and variable symbols defines a schema. Consider the string [01##], defined over the union of {#} and the alphabet $A = \{0,1\}$. This set includes [0100], [0101], [0110], and [0111]. Holland (1975, pp. 66–74) suggested that every string that is evaluated offers partial information about the expected fitness of all possible schemata in which that string resides. That is, if the string [0000] is evaluated to have some fitness, then partial information is also received about the worth of sampling from variations in [####], [0###], [#0##], [#00#], [#0#0], and so forth. This characteristic is termed *implicit parallelism* in that, through a single sample, infor-

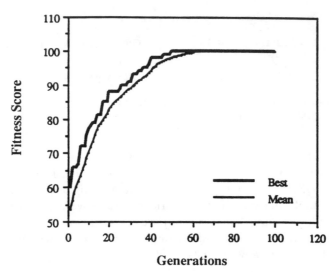

Figure 3-8 The results of a genetic algorithm maximizing the number of bits that are set equal to 1 in a 100-bit chromosome. Fitness is simply taken to be the number of 1s in the coding structure. The curves depict the fitness of the best chromosome in the population of 100 such chromosomes, as well as the mean fitness of all parents as a function of the number of generations.

mation is gained with respect to many schemata. Holland (1975, p. 71) speculated that it would be beneficial to maximize the number of schemata being sampled, thus providing maximum implicit parallelism, and he proved that this is achieved for $|A| = 2$. Binary strings were therefore suggested as a universal representation.

Currently, however, binary strings are rarely used for problems that are not obviously well mapped to a series of Boolean decisions or a bit mask. Much of the change in perspective has grown from practical experience: Michalewicz (1992, p. 82) indicated that for real-valued numerical optimization problems, floating-point representations outperform binary representations because they are more consistent, more precise, and lead to faster execution. Koza (1989), Davis (1991), Montana (1991), Syswerda (1991), Wright (1991), and others offered many similar observations and achieved useful results for difficult problems without relying on binary strings. More definitively, Fogel and Ghozeil (1997) proved that there are essential equivalencies between any bijective representations, regardless of cardinality (i.e., an algorithm with base two representation can be encoded exactly in, say, hexadecimal representation). Thus, no intrinsic advantage accrues to any particular representation.

Another issue to consider is that selection in proportion to fitness can be problematic. There are two practical considerations: (1) roulette wheel selection depends on positive values, and (2) the simple addition of a large constant value to the objective function can eliminate selection, with the algorithm then proceeding as a purely random walk. Several heuristics have been devised to compensate for these issues. For example, the fitness of all parents can be scaled relative to the mean fitness in the population (Goldberg, 1989, p. 77), or proportional selection can be based on ranking by fitness. Selection based on ranking also eliminates problems with functions that have large offsets.

One mathematical problem with selecting parents to reproduce in proportion to their relative fitness is that this procedure cannot ensure asymptotic convergence to a global optimum (Rudolph, 1994). The best chromosome in the population may be lost at any generation, and there is no assurance that any gains made up to a given generation will be retained in future generations. This can be overcome by employing a heuristic termed *elitist selection* (Grefenstette, 1986), which always retains the best chromosome in the population. This procedure guarantees asymptotic convergence (Eiben et al., 1991; Fogel, 1994; Rudolph, 1994), but the specific rates of convergence vary by problem and are generally unknown. Greater consideration to convergence rates is offered in Chapter 4.

From the mid-1980s to the early 1990s, genetic algorithms were distinguished by a strong emphasis on crossover, with bit mutation serving as a background operator to ensure that all possible alleles can enter the population (following Holland, 1975, pp. 110–111). The probabilities commonly assigned to crossover and bit mutation reflected this philosophical view. But several recent comparisons between systems that rely on crossover or mutation have not generally favored either strategy (Jones, 1995; Angeline, 1997; Chellapilla, 1997; Luke and Spector, 1998; Fuchs, 1998; and others). Moreover, when crossover is employed, the choice of which crossover operator to use is not straightforward.

Holland (1975, p. 160) and others (e.g., Goldberg, 1983; Grefenstette et al., 1985) proposed that genetic algorithms work by identifying good "building blocks" and eventually combining these to get larger building blocks. This idea has become known as the *building block hypothesis*. The hypothesis suggests that a one-point crossover operator would perform better than an operator that, say, took each bit from either parent with equal probability (uniform crossover), because it could maintain sequences (blocks) of "good code" that are associated with above-average performance and not disrupt their linkage. But this has not always been clearly demonstrated in the literature. Syswerda (1989) conducted function optimization experiments with uniform crossover, two-point crossover, and one-point crossover. Uniform crossover generally provided better solutions with less computational effort (see other related work in Reed et al., 1967, discussed below). Moreover, it has been noted that sections of code that reside at opposite ends of a chromosome are more likely to be disrupted under one-point crossover than are sections that are near the middle of the chromosome. Holland (1975, pp. 106–109) proposed an inversion operator that would reverse the index position for a section of the chromosome so that linkages could be constructed between arbitrary genes. Earlier, Fraser (1968) suggested this operator as well, but inversion has not been found to be useful in practice (Davis, 1991, p. 21). The relevance of the building block hypothesis is presently unclear, but its value is likely to vary significantly by problem.

Premature convergence has been another important concern in genetic algorithms. This problem occurs when the population of chromosomes reaches a configuration such that crossover no longer produces offspring that can outperform their parents, as must be the case in a homogeneous population. Under such circumstances, all standard forms of crossover simply regenerate the current parents. Any further optimization relies solely on bit mutation and can be quite slow. Premature convergence has often been observed in genetic algorithm research (De Jong, 1975; Pitney et al., 1990; Davis, 1991; Bickel and Bickel, 1991; and others) because of the exponential reproduction of the best observed chromosomes coupled with the strong emphasis on crossover. Davis (1991, pp. 26–27) recommended that when the population converged on a chromosome that would require the simultaneous mutation of many bits in order to improve it, the run was practically completed and it should either be restarted using a different random seed, or hill-climbing heuristics should be employed to search for improvements.

One interesting proposal for alleviating the problems associated with premature convergence was offered in Schraudolph and Belew (1992). The method, termed *dynamic parameter encoding* (DPE), dynamically resizes the available range of each parameter. Broadly, when a heuristic suggests that the population has converged, the minimum and maximum values for the range are resized to a smaller window and the process is iterated. In this manner, DPE can zoom in on solutions that are closer to the global optimum than provided by the initial precision. Schraudolph (personal communication) kindly provided results from experiments with DPE presented in Schraudolph and Belew (1992). As indicated in Figure 3-9, DPE outperforms the standard genetic algorithm when searching a quadratic bowl but performs worse on a multimodal function (Shekel's foxholes). The effectiveness of DPE is an open, promising area of research. Because DPE only zooms in, the initial range of parameters must be set to include the global optimum or it will not be found. But it would be relatively straightforward to include a mechanism in DPE to expand the search window as well as reduce it. Greater consideration to the details of DPE are offered in Chapter 4.

Although Fraser (1957a) and Holland (1975) proposed genetic algorithms primarily as models of genetic adaptive systems, Bremermann (1958; and others) emphasized the optimization characteristics of evolution. Bremermann (1962) considered the problem of minimizing a real-valued fitness function $f(x_1, \ldots , x_n)$, where $x_i \in R$. Simple functions were selected in order to pro-

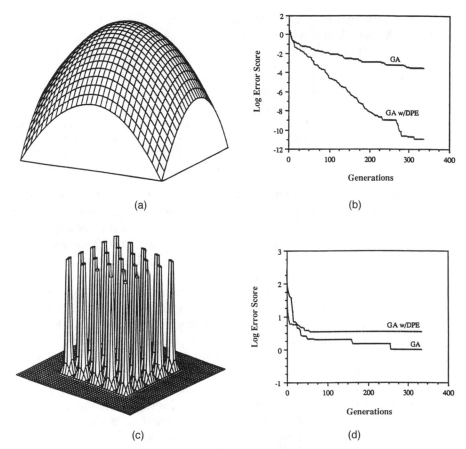

(a)   (b)

(c)   (d)

Figure 3-9 Results comparing genetic algorithms to genetic algorithms with dynamic parameter encoding. (a) An inverted quadratic function shown in two dimensions. (b) The results of optimization on a three-dimensional quadratic function. Dynamic parameter encoding provides a substantial advantage. (c) An inverted illustration of Shekel's foxholes. (d) The results of optimization on Shekel's foxholes. Dynamic parameter encoding does not appear to improve the efficiency or effectiveness of the genetic search on this function.

vide for a more tractable analysis of the evolutionary process. Linear systems of equations were considered, with the goal being to evolve a solution vector x that minimized $\|Ax - b\|_2$, where A is an $n \times n$ matrix and b is an $n \times 1$ column vector. The elements x_i were claimed to be analogous to an organism's genes. If only one of these elements was mutated at a time, the evolutionary process led quickly "to a stable point . . . [which] need not even be a saddle point [of the fitness function]" (Bremermann, 1962). For this evolutionary scheme to succeed, multiple elements had to be altered simultaneously, leading to a combinatorial explosion of possibilities. Bremermann (1962) suggested that although each element could be given a probability of changing, such a mechanism would be extremely slow to overcome stagnation.

Experiments indicated that:

> depending upon the step size we reach a point of stagnation which may or may not be near the solution. For well conditioned systems (ratio of the largest to smallest eigenvalue close to one) and with some previous knowledge about the range of variables and for n not large . . . the method gets us in a few minutes machine time close to the solution. For badly conditioned systems the process gets trapped at a point of stagnation. . . . A reduction in the step size results in further progress but after a while the process gets trapped again (Bremermann, 1962).

This was later termed the *trap phenomenon* (Bremermann and Rogson, 1964).

It appears that this trap phenomenon was subsequently misinterpreted. Ashby (1968) asserted that in general dynamic systems, as the system is made larger (i.e., with more components to be optimized), the response curve changes from a steady slope to a step function. But the response surface in question here remains a quadratic function regardless of the dimensionality. A gradient technique would find the globally optimal solution. The stagnation points that Bremermann encountered were not local optima, but rather points on the outside of elliptical contour lines of the multidimensional parabolic surface such that the fixed-length movement along a coordinate axis always resulted in a new point that resided on a higher contour ellipse (Figure 3-10).

Bremermann conjectured that sexual mating overcomes the situation where asexual evolution is stagnant. Mating schemes that combined elements of solutions in pairs and greater collections by linear superposition and by random combinations were tried. "None of these schemes, however, gave any spectacular results" (Bremermann, 1962). Attention was turned to linear programming problems, and again the results were undesirable. Random changes to only single components resulted in rapid stagnation. Furthermore, the combination of two independently discovered stagnation points ("sexual recombination") also stagnated on problems of modest size (eight variables).

Subsequent investigation (Bremermann and Rogson, 1964) indicated that evolving in random directions, rather than along the axes of the coordinate system, provided for a better search in linear programming problems. A fixed mutation size of high or low length was added to each component of the evolving solution vector. The superiority of this search scheme is to be expected. A much greater range of phenotypic expression is available to the evolving solution. Further improvements could have been expected if the step size had been made a continuous random variable (e.g., Gaussian). Extensions of this direction-based form of stochastic search with rotating coordinates are detailed in Wilde (1964, pp. 151–155; Wilde and Beightler, 1967, pp. 307–313).

About the same time as Bremermann's efforts, Reed et al. (1967) independently offered a comparable procedure that employed the idea of coevo-

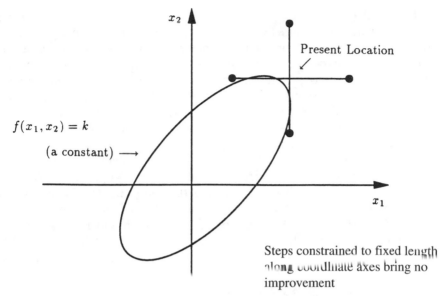

Figure 3-10 The ellipse represents a contour on the response surface being searched. The present
location is at a point that is higher than this contour, but no improvement can be
made by walking in a direction that is aligned with the coordinate axes for the specified fixed length. The step size could be reduced to attain an improvement, but an
equivalent ellipse would exist such that no further gains could be made with this reduced step.

lution to evaluate alternative strategies in a simplified game of poker. The
strategies were controlled by various parameters that underwent mutation, recombination, and selection. There were two players, and each received only
one card. There were two types of cards: high and low. Three types of bets were
allowed: pass, low bet, and high bet. A pass bet cost two units, a low bet cost
three units, and a high bet cost seven units. If the two opponents bet differently, the highest bet always won. If the bets were the same, the higher hand
won. In the event of a tie, no units were exchanged.

The betting strategies were defined in terms of the probabilities of betting "pass" with a low hand (L_P), betting "pass" with a high hand (H_P), betting "low" with a low hand (L_L), and betting "low" with a high hand (H_L).
The remaining probabilities are simply functions of these terms. In addition,
four mutational parameters were associated with each strategy (Reed et al.,
1967). These controlled the manner in which the strategies would be perturbed. Generating a new solution by crossing over was handled by selecting
each element from either parent with a 50 percent probability. Experiments
were conducted in which 50 parents were maintained at each generation. The
parents and offspring (one per parent) were evaluated over 20 trials of the
game.

Reed et al. (1967) concluded that "under conditions simulating polygenic control of quantitative characters, crossing does not enhance the speed of selective adaptation . . . the breeding characteristic spreads less rapidly or not at all." Additional experiments were conducted in which the effects of the genetic structure were nonpolygenic. Under these conditions, "crossing greatly enhances the speed of selective adaptation, particularly if interaction between the expressions of different hereditary factors is avoided" (Reed et al., 1967). As noted previously, however, it is precisely this interactivity among genetic effects (polygeny and pleiotropy) that predominates in living systems.

Although many open questions remain, genetic algorithms have been used to successfully address diverse practical optimization problems (Bäck et al., 1997). Current research efforts include the following. (1) Developing a stronger mathematical foundation for the genetic algorithm as an optimization technique (Davis and Principe, 1991; Schraudolph and Belew, 1992; Qi and Palmieri, 1993; Rudolph, 1994; Whitley and Vose, 1995; Salomon, 1996; Belew and Vose, 1997; also see early efforts to find the optimum mutation rate in Bremermann, 1958), including analysis of classes of problems that are difficult for genetic algorithms (Liepins and Vose, 1991; Whitley, 1991; Hart and Belew, 1991; Forrest and Mitchell, 1993), as well as sensitivity to performance of the general technique to various operator and parameter settings (Vignaux and Michalewicz, 1991; Michalewicz, 1992; Schaffer and Eshelman, 1991; Spears and De Jong, 1991; Kreinovich et al., 1993). (2) Comparing genetic algorithms to other optimization methods and examining the manner in which they can be enhanced by incorporating other procedures such as simulated annealing (Ingber and Rosen, 1992; Adler, 1993; Shapiro and Wu, 1996; Djurisic et al., 1997; Zacharias et al., 1998; Chen et al., 1998). (3) Using genetic algorithms for computer programming and engineering problems (Forrest and Mayer-Kress, 1991; Kristinsson and Dumont, 1992; Koza, 1992; and many others, see Bäck et al., 1997). (4) Applying genetic algorithms to rule-based classifier systems (Riolo, 1991; Liepins et al., 1991; Abu Zitar and Hassoun, 1993; Wilson, 1995). (5) Using genetic algorithms as a basis for artificial life simulations (Jefferson et al., 1991; Holland, 1992, pp. 186–195; Bicking et al., 1994; Mayoh, 1996). (6) Implementing genetic algorithms on parallel machines (Spiessens and Manderick, 1991, Mühlenbein et al., 1991; Matsumura et al., 1998). The most recent investigations can be found in Eshelman (1995) and Bäck (1997) (also see Forrest, 1993, and Belew and Booker, 1991).

3.6 THE EVOLUTION OF EVOLUTIONARY COMPUTATION

When the initial experiments in evolutionary computation were offered, they often met pointed criticism. For example, Minsky (1961) criticized the inability to rapidly discover suitable computer programs as evidenced in Friedberg

(1958; Friedberg et al., 1959), calling it "a comparable failure. . . . the machine did learn to solve some extremely simple problems. But it took on the order of 1,000 times longer than pure chance would expect." In retrospect, this assessment is an exaggeration; the worst case performance of Herman was 10 times longer than pure chance would expect. Moreover, Ramsy consistently outperformed random search. Minsky (1961) did agree that "in the last section of [Friedberg et al. (1959)] we see some real success obtained by breaking the problem into parts and solving them sequentially." Unfortunately, the relative successes of Friedberg et al. (1959) were overshadowed by the criticism of the earlier research (Friedberg, 1958).

Ramsy had a tendency to have all of its instructions bound (i.e., fixated) without being able to completely solve the problem at hand. In effect, the program had climbed to the top of a locally optimal hill without reaching the global optimum solution. Minsky attributed this problem to what he termed the "Mesa phenomena" (Minsky and Selfridge, 1961). "In changing just one instruction at a time the machine had not taken large enough steps in its search through the program space" (Minsky, 1961). This criticism would be appropriate for Herman, but not for Ramsy, which altered all of its unbound program instructions after every failure.

One important limitation of the early efforts in evolutionary computation was the available computing speed. Even the large mainframe computers in the early 1960s were about one-half the speed of an Apple II. Thus, the ideas were often far ahead of the technology that was required to make them successful. Shortcuts were often mandatory, such as always focusing attention on the absolute best available individuals in a population, as opposed to giving more attention to lesser-valued individuals. Pincus (1970) and later Galar (1985, 1989) analyzed the case where less-fit individuals could be saved as parents for subsequent generations. Through a series of simple experiments, Galar demonstrated that "the escapes from evolutionary traps are due to mutations of handicapped individua of intermediary types." That is, it is important to maintain a probabilistic selection mechanism, or at the least, less-fit solutions at each generation.

Galar (1991) examined the ability of what was described as a "sexually reproducing" small population to overcome saddles in the adaptive landscape. The recombination operation involved differs from that utilized in typical genetic algorithm research. Galar (1991) defined an individual as a set of real-valued "traits." In asexual reproduction, the value of the jth virtual trait of the kth individual in the population as a descendant from the rth individual is given by

$$x(t)_{k,j} = x(t - 1)_{r,j} + e,$$

where t represents the generation number and e represents a Gaussian random variable with mean zero and constant variance. In sexual reproduction, the value of the jth virtual trait of the kth individual in the population, which is a descendant of the rth couple, is given by

$$x(t)_{k,j} = 0.5(x(t-1)_{2r-1,j} + x(t-1)_{2r,j}) + e.$$

Thus, the sexual reproduction being studied was actually a "blending" of the traits. Galar (1991) noted that the "transitions [across saddles in the adaptive landscape] of sexual populations are much more difficult [than for asexual populations]. Small populations can cross saddles, but this ability fades with population growth. . . . The main factor is reproduction of the midparental values of traits: even if some individual gets across a saddle, it usually mates with a partner who is still close to the old local optimum. Because of this, randomly mating populations have to cross saddles as a whole." Galar (1991) concluded that the "resulting differences in the waiting time for saddle crossing can be immense, even if the population consists of [only a small number of] individuals."

Another idea that had to await faster processing machinery was that of evolving neural networks. This notion was originally discussed by Bremermann and Frank Rosenblatt at Cornell in the early 1960s (Bremermann, 1968, 1996), as evolutionary optimization might be used to train the perceptron. Klopf (1965) and Klopf and Gose (1969) offered some simple experiments with an "evolutionary" technique that operated on a single neural network, but perhaps the first modern effort to apply evolutionary algorithms to optimizing neural networks came in Kampfer and Conrad (1983). Earlier, Conrad (1974) proposed an "evolutionary learning circuits model" in which he hypothesized that the brain harnesses the same types of mechanisms for learning as does evolution. Kampfer and Conrad (1983) simulated neuronal learning systems that adapt through evolutionary changes in which the time required to learn various pattern recognition tasks increased as a polynomial function of the size of the task.

Prior to the early 1990s, efforts in evolutionary computation had progressed mainly along the three separate tracks of evolutionary programming, evolution strategies, and genetic algorithms. Each of these efforts had adopted slightly different perspectives of modeling evolution at the level of species, individuals, and genes, respectively. Given the obvious similarities of the approaches, there was surprisingly little contact between the communities. In 1993, the term *evolutionary computation* was developed for the first journal that specifically covered efforts in all of these areas. This served as a rubric to foster interaction between these groups. The term was quickly picked up by several other organizations holding conferences (e.g., the IEEE) and has become the term of the art. The effect of emphasizing the commonalties between the different lines of investigating simulated evolution has been profound.

Within the past five years, each area of evolutionary computation has borrowed and modified ideas from the others. Over time, an iterative blending has occurred such that all classes of evolutionary algorithms now appear quite similar, if not for all intents and purposes identical. It is no longer possible to identify a particular effort in evolutionary computation as a genetic algorithm, an evolution strategy, or an evolutionary program, simply by examining the representation chosen, the selection method, the use of self-adaptation, recombination, or any other factor. In fact, the practical utility of each of these terms has evolved to be essentially useless: Little or no information is conveyed by identifying a particular effort as a genetic algorithm, evolution strategy, or evolutionary program. Any of these efforts can use arbitrary data structures, a variety of selection techniques ranging from (weak) proportional selection to (strong) truncation selection, a diverse array of variation operators, including those that use single parents or multiple parents, and so forth. There are no theoretical results for, say, evolution strategies, that are separate from theory in evolutionary computation more broadly. Moreover, several early conjectures about the optimality of maximizing schema processing and sampling strategies that were offered within genetic algorithm research have recently been shown mathematically to be flawed (see Chapter 4 for details), providing even less impetus for differentiating between "genetic" and other evolutionary algorithms. The most rapid progress in evolutionary computation can come not by treating these communities as "separate but equal" demes, but rather by eschewing efforts to separate them at all.

The essential ingredients in any evolutionary algorithm are population-based random variation and selection. Regardless of the particular implementation, no model of evolution can be a complete description of a true, natural evolutionary system. Every evolutionary simulation is incomplete. But these simulations have also been demonstrated to be of practical use when applied to difficult optimization problems. The greatest potential for the application of evolutionary optimization to real-world problems will come from their implementation on parallel machines, for evolution is an inherently parallel process. Each parent can be varied independently, and each offspring can be evaluated independently (unless fitness is an explicit function of other members of the population, e.g., Fogel, 1993c, and even then the process can be accelerated through parallelism). Competition and selection can also be parallelized. Recent advances in distributed and parallel processing architectures will result in dramatically reduced execution times for simulations that would simply be impractical on current serial computers.

Natural evolution is a most robust yet efficient problem-solving technique. Evolutionary computation can likewise be made robust. The same procedures can be applied to diverse problems with relatively little reprogramming. Evolutionary systems, be they natural or simulated, are intelligent under the definition proposed in Chapter 1. They adapt their behavior to meet goals in a range of environments. The property of intelligence is not the "end-point of evolutionary search. Rather, intelligence is inseparable from the trial-and-error process itself" (Atmar, 1990).

REFERENCES

Abu Zitar, R. A., and M. H. Hassoun (1993). "Regulator Control via Genetic Search Assisted Reinforcement," In *Proc. of the Fifth Intern. Conf. on Genetic Algorithms,* edited by S. Forrest, San Mateo, CA: Morgan Kaufmann, pp. 254–262.

Adler, D. (1993). "Genetic Algorithms and Simulated Annealing: A Marriage Proposal," In *IEEE Conference on Neural Networks 1993,* Piscataway, NJ: IEEE Press, pp. 1104–1109.

Akaike, H. (1974). "A New Look at the Statistical Model Identification," *IEEE Trans. Auto. Control,* Vol. 19:6, pp. 716–723.

Ambati, B. K., J. Ambati, and M. M. Mokhtar (1991). "Heuristic Combinatorial Optimization by Simulated Darwinian Evolution: A Polynomial Time Algorithm for the Traveling Salesman Problem," *Biological Cybernetics,* Vol. 65, pp. 31–35.

Angeline, P. J. (1996). "The Effects of Noise on Self-Adaptive Evolutionary Optimization," In *Evolutionary Programming V: Proc. 5th Annual Conference on Evolutionary Programming,* edited by L. J. Fogel, P. J. Angeline, and T. Bäck, Cambridge, MA: MIT Press, pp. 433–439.

Angeline, P. J. (1997). "Subtree Crossover: Building Block Engine or Macromutation?" In *Genetic Programming 1997: Proceedings of the Second Annual Conference,* edited by J. R. Koza, K. Deb, M. Dorigo, D. B. Fogel, M. Garzon, H. Iba, and R. L. Riolo (eds.), San Francisco: Morgan Kaufmann, pp. 9–17.

Angeline, P. J., R. G. Reynolds, J. R. McDonnell, and R. Eberhart (eds.) (1997). *Evolutionary Programming VI,* Berlin: Springer.

Angeline, P. J., G. M. Saunders and J. B. Pollack (1994). "An Evolutionary Algorithm That Constructs Recurrent Neural Networks," *IEEE Trans. Neural Networks,* Vol. 5:1, pp. 54–65.

Antonisse, J. (1989). "A New Interpretation of Schema Notation That Overturns the Binary Encoding Constraint," In *Proc. of the Third Intern. Conf. on Genetic Algorithms,* edited by J. D. Schaffer, San Mateo, CA: Morgan Kaufmann, pp. 86–91.

Ashby, W. R. (1968). "Contribution of Information Theory to Pathological Mechanisms," *British J. of Psychiatry,* Vol. 114, pp. 1485–1498.

Atmar, J. W. (1976). "Speculation on the Evolution of Intelligence and Its Possible Realization in Machine Form," Sc.D. Diss., New Mexico State University, Las Cruces.

Atmar, W. (1990). "Natural Processes Which Accelerate the Evolutionary Search," In *Proc. of the 24th Asilomar Conf. on Signals, Systems, and Computers,* edited by R. R. Chen, Pacific Grove, CA: Maple Press, pp. 1030–1035.

Bäck, T. (ed.) (1997). *Proc. 7th Intern. Conference on Genetic Algorithms,* San Francisco: Morgan Kaufmann.

Bäck, T., D. B. Fogel, and Z. Michalewicz (eds.) (1997). *Handbook of Evolutionary Computation,* New York: Oxford/IOP.

Bäck, T., F. Hoffmeister, and H.-P. Schwefel (1991). "A Survey of Evolution Strategies," In *Proc. of the Fourth Intern. Conf. on Genetic Algorithms,* edited by R. K. Belew and L. B. Booker, San Mateo, CA: Morgan Kaufmann, pp. 2–9.

Bäck, T., G. Rudolph, and H.-P. Schwefel (1993). "Evolutionary Programming and Evolution Strategies: Similarities and Differences," In *Proc. of the Second Ann. Conf. on Evolutionary Programming,* edited by D. B. Fogel and W. Atmar, La Jolla, CA: Evolutionary Programming Society, pp. 11–22.

Bäck, T., and H.-P. Schwefel (1993). "An Overview of Evolutionary Algorithms for Parameter Optimization," *Evolutionary Computation,* Vol. 1:1, pp. 1–23.

Bagley, J. D. (1967). "The Behavior of Adaptive Systems Which Employ Genetic and Correlation Algorithms," Doctoral Dissertation, Univ. Michigan, Ann Arbor.

Barnett, E. H. (1960). "Introduction to EVOP," *Industr. and Eng. Chem.,* Vol. 52:6, pp. 500–503.

Barricelli, N. A. (1954). "Esempi Numerici di Processi di Evoluzione," *Methodos,* pp. 45–68.

Barricelli, N. A. (1957). "Symbiogenetic Evolution Processes Realized by Artificial Methods," *Methodos,* Vol. 9:35–36, pp. 143–182.

Barricelli, N. A. (1962). "Numerical Testing of Evolution Theories: I. Theoretical Introduction and Basic Tests," *Acta Biotheoretica,* Vol. 16:1–2, pp. 69–98.

Barricelli, N. A. (1963). "Numerical Testing of Evolution Theories: II. Preliminary Tests of Performance Symbiogenesis and Terrestrial Life," *Acta Biotheoretica,* Vol. 16:3–4, pp. 99–126.

Barron, A. R. (1984). "Predicted Squared Error: A Criterion of Automatic Model Selection," In *Self-Organizing Methods in Modeling,* edited by S. J. Farlow, New York: Marcel Dekker, pp. 87–103.

Belew, R. K., and L. B. Booker (eds.) (1991). *Proc. 4th Intern. Conf. on Genetic Algorithms,* San Mateo, CA: Morgan Kaufmann.

Belew, R. K., and M. D. Vose (eds.) (1997). *Foundations of Genetic Algorithms 4,* San Mateo, CA: Morgan Kaufmann.

Beyer, H.-G. (1997). "An Alternative Explanation for the Manner in Which Genetic Algorithms Operate," *BioSystems,* Vol. 41:1, pp. 1–15.

Bickel, A. S., and R. W. Bickel (1990). "Determination of Near-Optimum Use of Hospital Diagnostic Resources Using the 'GENES' Genetic Algorithm," *Comput. Biol. Med.,* Vol. 20:1, pp. 1–13.

Bicking, F., C. Fonteix, J.-P. Corriou, and I. Marc (1994). "Global Optimization by Artificial Life: A New Technique Using Genetic Population Evolution," *RAIRO Recherche Operationelle,* Vol. 28:1, pp. 23–36.

Born, J., H.-M. Voigt, and I. Santibanez-Koref (1992). "Alternative Evolution Strategies to Global Optimization," In *Parallel Problem Solving from Nature 2,* edited by R. Männer and B. Manderick, Amsterdam: North-Holland, pp. 187–196.

Box, G. E. P. (1957). "Evolutionary Operation: A Method for Increasing Industrial Productivity," *Applied Statistics,* Vol. 6, pp. 81–101.

Box, G. E. P. (1960). "Some General Considerations in Process Optimization," *J. Basic Eng.,* March, pp. 113–119.

Box, G. E. P. (1966). "A Simple System of Evolutionary Operation Subject to Empirical Feedback," *Technometrics,* Vol. 8, pp. 19–26.

Box, G. E. P., and J. Chanmugam (1962). "Adaptive Optimization of Continuous Processes," *Indust. and Eng. Chem. Fundamentals,* Vol. 1, pp. 2–16.

Box, G. E. P., and N. R. Draper (1969). *Evolutionary Operation: A Statistical Method for Process Control,* New York: John Wiley.

Box, G. E. P., and J. S. Hunter (1959). "Condensed Calculations for Evolutionary Operation Programs," *Technometrics,* Vol. 1, pp. 77–95.

Bremermann, H. J. (1958). "The Evolution of Intelligence. The Nervous System as a Model of Its Environment," Technical Report No. 1, Contract No. 477(17), Dept. of Mathematics, Univ. of Wash., Seattle, July.

Bremermann, H. J. (1962). "Optimization Through Evolution and Recombination," In *Self-Organizing Systems,* edited by M. C. Yovits, G. T. Jacobi, and G. D. Goldstine, Washington, DC: Spartan Books, pp. 93–106.

Bremermann, H. J. (1967). "Quantitative Aspects of Goal-Seeking Self-Organizing Systems," *Progress in Theoretical Biology,* Vol. 1, New York: Academic Press, pp. 59–77.

Bremermann, H. J. (1968). "Numerical Optimization Procedures Derived from Biological Evolution Processes," In *Cybernetic Problems in Bionics,* edited by H. L. Oestreicher and D. R. Moore, New York: Gordon and Breach, pp. 543–562.

Bremermann, H. J. (1973). "On the Dynamics and Trajectories of Evolution Processes," In *Biogenesis, Evolution, Homeostasis,* edited by A. Locker, New York: Springer-Verlag, pp. 29–37.

Bremermann, H. J. (1996). Personal Communication, U.C. Berkeley.

Bremermann, H. J., and M. Rogson (1964). "An Evolution-Type Search Method for Convex Sets," ONR Tech. Report, Contracts 222(85) and 3656(58), Berkeley, CA, May.

Bremermann, H. J., M. Rogson, and S. Salaff (1965). "Search by Evolution," In *Biophysics and Cybernetic Systems,* edited by M. Maxfield, A. Callahan, and L. J. Fogel, Washington, DC: Spartan Books, pp. 157–167.

Bremermann, H. J., M. Rogson, and S. Salaff (1966). "Global Properties of Evolution Processes," In *Natural Automata and Useful Simulations,* edited by H. H. Pattee, E. A. Edlsack, L. Fein, and A. B. Callahan, Washington, DC: Spartan Books, pp. 3–41.

Brody, W. M., and N. Lindgren (1968). "Human Enhancement: Beyond the Machine Age," *IEEE Spectrum,* February, pp. 79–93.

Burgin, G. H. (1969). "On Playing Two-Person Zero-Sum Games Against Nonminimax Players," *IEEE Trans. on Systems Science and Cybernetics,* Vol. SSC-5:4, pp. 369–370.

Burgin, G. H. (1974). "System Identification by Quasilinearization and Evolutionary Programming," *Journal of Cybernetics,* Vol. 2:3, pp. 4–23.

Cai, J. B., and G. Thierauf (1996). "Evolution Strategies for Solving Discrete Optimization Problems," *Adv. Engin. Software,* Vol. 25:2–3, pp. 177–183.

Campbell, D. T. (1956a). "Adaptive Behavior from Random Response," *Behavioral Science,* Vol. 1, pp. 105–110.

Campbell, D. T. (1956b). "Perception as Substitute Trial and Error," *Psychol. Rev.,* Vol. 63, pp. 330–342.

Campbell, D. T. (1960). "Blind Variation and Selective Survival as a General Strategy in Knowledge-Processes," In *Self-Organizing Systems,* edited by M. C. Yovits and S. Cameron, New York: Pergamon Press, pp. 205–231.

Cannon, W. D. (1932). *The Wisdom of the Body,* New York: W. W. Norton.

Chandrasekaran, B., and L. H. Recher (1974). "Artificial Intelligence—A Case for Agnosticism," *IEEE Trans. on Sys., Man and Cyber.,* Vol. SMC-4:1, pp. 88–94.

Chellapilla, K. (1997). "Evolving Computer Programs Without Subtree Crossover," *IEEE Trans. Evolutionary Computation,* Vol. 1:3, pp. 209–216.

Chen, H., N. S. Flann, and D. W. Watson (1998). "Parallel Genetic Simulated Annealing: A Massively Parallel SIMD Algorithm," *IEEE Trans. Parallel and Distributed Systems,* Vol. 9:2, pp. 126–136.

Conrad, M. (1974). "Evolutionary Learning Circuits," *J. Theo. Biol.,* Vol. 46, pp. 167–188.

Conrad, M. (1981). "Algorithmic Specification as a Technique for Computing with Informal Biological Models," *BioSystems,* Vol. 13, pp. 303–320.

Conrad, M., and H. H. Pattee (1970). "Evolution Experiments with an Artificial Ecosystem," *J. Theoret. Biol.,* Vol. 28, pp. 393–409.

Conrad, M., and M. M. Rizki (1989). "The Artificial Worlds Approach to Emergent Evolution," *BioSystems,* Vol. 23, pp. 247–260.

Cornett, F. N. (1972). "An Application of Evolutionary Programming to Pattern Recognition," Master's Thesis, New Mexico State University, Las Cruces, NM.

Davidor, Y., H.-P. Schwefel, and R. Männer (eds.) (1994). *Parallel Problem Solving from Nature 3,* Berlin: Springer.

Davis, L. (ed.) (1991). *Handbook of Genetic Algorithms,* New York: Van Nostrand Reinhold.

Davis, T. E., and J. C. Principe (1991). "A Simulated Annealing Like Convergence Theory for the Simple Genetic Algorithm," In *Proc. of the Fourth Intern. Conf. on Genetic Algorithms,* edited by R. K. Belew and L. B. Booker, San Mateo, CA: Morgan Kaufmann, pp. 174–181.

Dearholt, D. W. (1976). "Some Experiments on Generalization Using Evolving Automata," In *Proc. 9th Inter. Conf. on System Sciences,* Honolulu, HI, pp. 131–133.

De Jong, K. A. (1975). "The Analysis of the Behavior of a Class of Genetic Adaptive Systems," Doctoral Dissertation, Univ. of Michigan, Ann Arbor.

Djurisic, A. B., J. M. Elazar, and A. D. Rakic (1997). "Simulated-Annealing-Based Genetic Algorithm for Modeling the Optical Constants of Solids," *Applied Optics,* Vol. 36: 28, pp. 7097–7103.

Duncan, B. S. (1993). "Parallel Evolutionary Programming," In *Proc. of the Second Annual Conf. on Evolutionary Programming,* edited by D. B. Fogel and W. Atmar, La Jolla, CA: Evolutionary Programming Society, pp. 202–208.

Dunham, B., D. Fridshal, R. Fridshal, and J. H. North (1963). "Design by Natural Selection," *Synthese,* Vol. 15, pp. 254–259.

Eiben, A. E., E. H. Aarts, and K. M. Van Hee (1991). "Global Convergence of Genetic Algorithms: An Infinite Markov Chain Analysis," In *Parallel Problem Solving from Nature,* edited by H.-P. Schwefel and R. Männer, Berlin: Springer, pp. 4–12.

Eiben, A. E., M. Schoenauer, T. Bäck, and H.-P. Schwefel (eds.) (1998). *Parallel Problem Solving from Nature—PPSN V,* Berlin: Springer.

Eshelman, L. (ed.) (1995). *Proc. 6th Intern. Conf. on Genetic Algorithms,* San Mateo, CA: Morgan Kaufmann.

Fathi-Torbaghan, M., and L. Hildebrand (1997). "Model-Free Optimization of Fuzzy Rule-Based Systems Using Evolution Strategies," *IEEE Trans. Sys., Man and Cybern. Part B.,* Vol. 27:2, pp. 270–277.

Feigenbaum, E. A. (1963). "Artificial Intelligence Research," *IEEE Trans. on Information Theory,* Vol. IT-9, pp. 248–253.

Flood, M. M. (1962). "Stochastic Learning Theory Applied to Chance Experiments with Cats, Dogs and Men," *Behavioral Science,* Vol. 7:3, pp. 289–314.

Fogel, D. B. (1988). "An Evolutionary Approach to the Traveling Salesman Problem," *Biological Cybernetics,* Vol. 60:2, pp. 139–144.

Fogel, D. B. (1991a). *System Identification Through Simulated Evolution: A Machine Learning Approach to Modeling,* Needham, MA: Ginn Press.

Fogel, D. B. (1991b). "The Evolution of Intelligent Decision-Making in Gaming," *Cybernetics and Systems,* Vol. 22, pp. 223–236.

Fogel, D. B. (1992). "Evolving Artificial Intelligence," Doctoral Dissertation, UCSD.

Fogel, D. B. (1993a). "Empirical Estimation of the Computation Required to Discover Approximate Solutions to the Traveling Salesman Problem Using Evolutionary Programming," In *Proc. of the Second Annual Conference on Evolutionary Programming,* edited by D. B. Fogel and W. Atmar (eds.), La Jolla, CA: Evolutionary Programming Society, pp. 56–61.

Fogel, D. B. (1993b). "Applying Evolutionary Programming to Selected Traveling Salesman Problems," *Cybernetics and Systems,* Vol. 24, pp. 27–36.

Fogel, D. B. (1993c). "Evolving Behaviors in the Iterated Prisoner's Dilemma," *Evolutionary Computation,* Vol. 1:1, pp. 77–97.

Fogel, D. B. (1994). "Asymptotic Convergence Properties of Genetic Algorithms and Evolutionary Programming: Analysis and Experiments," *Cybernetics and Systems,* Vol. 25:3, pp. 389–407.

Fogel, D. B. (1995). *Evolutionary Computation: Toward a New Philosophy of Machine Intelligence,* 1st ed., Piscataway, NJ: IEEE Press.

Fogel, D. B. (ed.) (1998). *Evolutionary Computation: The Fossil Record,* Piscataway, NJ: IEEE Press.

Fogel, D. B., and J. W. Atmar (1990). "Comparing Genetic Operators with Gaussian Mutations in Simulated Evolutionary Processes Using Linear Systems," *Biological Cybernetics,* Vol. 63, pp. 111–114.

Fogel, D. B., and W. Atmar (eds.) (1992). *Proceedings of the First Annual Conference on Evolutionary Programming,* La Jolla, CA: Evolutionary Programming Society.

Fogel, D. B., and W. Atmar (eds.) (1993). *Proceedings of the Second Annual Conference on Evolutionary Programming,* La Jolla, CA: Evolutionary Programming Society.

Fogel, D. B., and L. J. Fogel (1990). "Optimal Routing of Multiple Autonomous Underwater Vehicles through Evolutionary Programming," In *Proc. of the Symp. on Auton. Underwater Vehicle Tech.,* chaired by C. E. Stuart, Washington, DC, pp. 44–47.

Fogel, D. B., L. J. Fogel, and J. W. Atmar (1991). "Meta-Evolutionary Programming," In *Proc. of the 24th Asilomar Conf. on Signals, Systems and Computers,* edited by R. R. Chen, Pacific Grove, CA: IEEE Comp. Soc. Press, pp. 540–545.

Fogel, D. B., L. J. Fogel, W. Atmar, and G. B. Fogel (1992). "Hierarchic Methods of Evolutionary Programming," In *Proc. of the First Ann. Conf. on Evolutionary Programming,* edited by D. B. Fogel and W. Atmar, La Jolla, CA: Evolutionary Programming Society, pp. 175–182.

Fogel, D. B., L. J. Fogel, and V. W. Porto (1990). "Evolving Neural Networks," *Biological Cybernetics,* Vol. 63:6, pp. 487–493.

Fogel, D. B., and A. Ghozeil (1997). "A Note on Representations and Variation Operators," *IEEE Trans. Evolutionary Computation,* Vol. 1:2, pp. 159–161.

Fogel, D. B., and L. C. Stayton (1994). "On the Effectiveness of Crossover in Simulated Evolutionary Optimization," *BioSystems,* Vol. 32:3, pp. 171–182.

Fogel, D. B., E. C. Wasson, E. M. Boughton, V. W. Porto, and P. J. Angeline (1998). "Linear and Neural Models for Classifying Breast Cancer," *IEEE Trans. Medical Imaging,* Vol. 17:3, pp. 485–488.

Fogel, L. J. (1962a). "Autonomous Automata," *Industrial Research,* Vol. 4, pp. 14–19.

Fogel, L. J. (1962b). "Toward Inductive Inference Automata," *Proc. of the Int. Federation for Information Processing Congress,* Munich, pp. 395–400.

Fogel, L. J. (1962c). "Decision-Making by Automata," Technical Report GDA-ERR-AN-222, General Dynamics, San Diego.

Fogel, L. J. (1963). *Biotechnology: Concepts and Applications,* Englewood Cliffs, NJ: Prentice-Hall.

Fogel, L. J. (1964). "On the Organization of Intellect," Doctoral Dissertation, UCLA.

Fogel, L. J. (1966). "On the Design of Conscious Automata," Final Report under Contract No. AF 49(638)-1651, AFOSR, Arlington, VA.

Fogel, L. J. (1968). "Extending Communication and Control through Simulated Evolution," In *Bioengineering—An Engineering View, Proc. Symp. Engineering Significance of the Biological Sciences,* edited by G. Bugliarello, San Francisco: San Francisco Press, pp. 286–304.

Fogel, L. J., P. J. Angeline, and T. Bäck (eds.) (1996). *Evolutionary Programming V: Proceedings of the Fifth Annual Conference on Evolutionary Programming,* Cambridge, MA: MIT Press.

Fogel, L. J., and G. H. Burgin (1969). "Competitive Goal-Seeking Through Evolutionary Programming," Final Report under Contract No. AF 19(628)-5927, Air Force Cambridge Research Labs.

Fogel, L. J., A. J. Owens, and M. J. Walsh (1964). "On the Evolution of Artificial Intelligence," In *Proc. of the 5th National Symp. on Human Factors in Electronics,* San Diego, CA: IEEE, pp. 63–76.

Fogel, L. J., A. J. Owens, and M. J. Walsh (1965a). "Artificial Intelligence Through a Simulation of Evolution," In *Biophysics and Cybernetics Systems,* edited by A. Callahan, M. Maxfield, and L. J. Fogel, Washington, DC: Spartan Books, pp. 131–156.

Fogel, L. J., A. J. Owens, and M. J. Walsh (1965b). "Intelligent Decision-Making Through a Simulation of Evolution," *IEEE Trans. of the Professional Technical Group on Human Factors in Electronics,* Vol. 6:1.

Fogel, L. J., A. J. Owens, and M. J. Walsh (1966). *Artificial Intelligence Through Simulated Evolution,* New York: John Wiley.

Fogel, L. J., and M. J. Walsh (1964). "Research Toward Automatic Pattern Recognition and Artificial Intelligence," Technical Report GDR-ERR-AN-568, General Dynamics, San Diego.

Forrest, S. (ed.) (1993). *Proc. of the Fifth International Conference on Genetic Algorithms,* San Mateo, CA: Morgan Kaufmann.

Forrest, S., and G. Mayer-Kress (1991). "Genetic Algorithms, Nonlinear Dynamical Systems, and Models of International Security," In *Handbook of Genetic Algorithms,* edited by L. Davis, New York: Van Nostrand Reinhold, pp. 166–185.

Forrest, S., and M. Mitchell (1993). "What Makes a Problem Hard for a Genetic Algorithm—Some Anomolous Results and Their Explanation," *Machine Learning,* Vol. 13:2–3, pp. 285–319.

Franzen, O., H. Blume, and H. Schroder (1998). "FIR-Filter Design with Spatial and Frequency Design Constraints using Evolution Strategies," *Signal Processing,* Vol. 68:3, pp. 295–306.

Fraser, A. S. (1957a). "Simulation of Genetic Systems by Automatic Digital Computers. I. Introduction," *Australian J. of Biol. Sci.,* Vol. 10, pp. 484–491.

Fraser, A. S. (1957b). "Simulation of Genetic Systems by Automatic Digital Computers. II. Effects of Linkage on Rates of Advance under Selection," *Australian J. of Biol. Sci.,* Vol. 10, pp. 492–499.

Fraser, A. S. (1960). "Simulation of Genetic Systems by Automatic Digital Computers. VI. Epistasis," *Australian J. of Biol. Sci.,* Vol. 13, pp. 150–162.

Fraser, A. S. (1962). "Simulation of Genetic Systems," *J. of Theoretical Biology,* Vol. 2, pp. 329–346.

Fraser, A. S. (1967). "Comments on Mathematical Challenges to the Neo-Darwinian Concept of Evolution," In *Mathematical Challenges to the Neo-Darwinian Interpretation of Evolution,* edited by P. S. Moorhead and M. M. Kaplan, Philadelphia: Wistar Institute Press, p. 107.

Fraser, A. S. (1968). "The Evolution of Purposive Behavior," In *Purposive Systems,* edited by H. von Foerster, J. D. White, L. J. Peterson, and J. K. Russell, Washington, DC: Spartan Books, pp. 15–23.

Friedberg, R. M. (1958). "A Learning Machine: Part I," *IBM J. of Research and Development,* Vol. 2, pp. 2–13.

Friedberg, R. M., B. Dunham, and J. H. North (1959). "A Learning Machine: Part II," *IBM J. of Research and Development,* Vol. 3, pp. 282–287.

Friedman, G. J. (1956). "Selective Feedback Computers for Engineering Synthesis and Nervous System Analogy," Master's Thesis, UCLA.

Friedman, G. J. (1959). "Digital Simulation of an Evolutionary Process," *General Systems: Yearbook of the Society for General Systems Research,* Vol. 4, pp. 171–184.

Fuchs, M. (1998). "Crossover versus Mutation: An Empirical and Theoretical Case Study," In *Genetic Programming 98: Proceedings of the Third Annual Genetic*

Programming Conference, edited by J. R. Koza, W. Banzhaf, K. Chellapilla, K. Deb, M. Dorigo, D. B. Fogel, M. H. Garzon, D. E. Goldberg, H. Iba, and R. L. Riolo, San Francisco: Morgan Kaufmann, pp. 78–85.

Galar, R. (1985). "Handicapped Individua in Evolutionary Processes," *Biological Cybernetics,* Vol. 53, pp. 1–9.

Galar, R. (1989). "Evolutionary Search with Soft Selection," *Biological Cybernetics,* Vol. 60, pp. 357–364.

Galar, R. (1991). "Simulation of Local Evolutionary Dynamics of Small Populations," *Biological Cybernetics,* Vol. 65, pp. 37–45.

Goldberg, D. E. (1983). "Computer-Aided Gas Pipeline Operation Using Genetic Algorithms and Rule Learning," Doctoral Dissertation, Univ. of Michigan, Ann Arbor.

Goldberg, D. E. (1989). *Genetic Algorithms in Search, Optimization and Machine Learning,* Reading, MA: Addison-Wesley.

Grefenstette, J. J. (1986). "Optimization of Control Parameters for Genetic Algorithms," *IEEE Trans. Sys., Man and Cybern.,* Vol. SMC-16:1, pp. 122–128.

Grefenstette, J. J., R. Gopal, B. Rosmaita, and D. Van Gucht (1985). "Genetic Algorithms for the Traveling Salesman Problem," In *Proc. of an Intern. Conf. on Genetic Algorithms and Their Applications,* edited by J. J. Grefenstette, Hillsdale, NJ: Lawrence Erlbaum, pp. 160–168.

Gross, B., U. Hammel, P. Maldaner, A. Meyer, P. Roosen, and M. Schütz (1996). "Optimization of Heat Exchanger Networks by Means of Evolution Strategies," In *Parallel Problem Solving from Nature IV,* edited by H.-M. Voigt, W. Ebeling, I. Rechenberg, and H.-P. Schwefel, Berlin: Springer, pp. 1002–1011.

Hart, W. E., and R. K. Belew (1991). "Optimizing an Arbitrary Function Is Hard for the Genetic Algorithm," In *Proc. of the Fourth Intern. Conf. on Genetic Algorithms,* edited by R. K. Belew and L. B. Booker, San Mateo, CA: Morgan Kaufmann, pp. 190–195.

Hoffmeister, F. (1991). "Scalable Parallelism by Evolutionary Algorithms," In *Parallel Comp. & Math. Opt.,* edited by D. B. Grauer, Berlin: Springer, pp. 177–198.

Holland, J. H. (1969). "Adaptive Plans Optimal for Payoff-Only Environments," In *Proc. of the 2nd Hawaii Int. Conf. on System Sciences,* pp. 917–920.

Holland, J. H. (1973). "Genetic Algorithms and the Optimal Allocation of Trials," *SIAM J. Comp.,* Vol. 2:2, pp. 88–105.

Holland, J. H. (1975). *Adaptation in Natural and Artificial Systems,* Ann Arbor, MI: Univ. of Michigan Press.

Holland, J. H. (1992). *Adaptation in Natural and Artificial Systems,* 2nd ed, Cambridge, MA: MIT Press.

Holmes, V. P. (1973). "Recognizing Prime Numbers with an Evolutionary Program," Master's Thesis, New Mexico State University, Las Cruces, NM.

Hunter, J. S. (1958). "Some Statistical Principles Underlying EVOP," *Sec. Stevens Symp.,* pp. 63–75.

Hunter, J. S. (1960). "Optimize Your Process with EVOP," *Chem. Eng.,* Vol. 67, pp. 193–202.

Ingber, L., and B. Rosen (1992). "Genetic Algorithms and Very Fast Simulated Annealing—A Comparison," *Math. and Comp. Model.,* Vol. 16:11, pp. 87–100.

Jackson, P. C. (1974). *Introduction to Artificial Intelligence,* (reprint) New York: Dover Publications.

Jefferson, D., R. Collins, C. Cooper, M. Dyer, M. Flowers, R. Korf, C. Taylor, and A. Wang (1991). "Evolution as a Theme in Artificial Life: The Genesys/Tracker System," In *Artificial Life II,* edited by C. G. Langton, C. Taylor, J. D. Farmer, and S. Rasmussen, Reading, MA: Addison-Wesley, pp. 549–578.

Jones, T. (1995). "Crossover, Macromutation, and Population-Based Search," In *Proc. of 6th Intern. Conf. on Genetic Algorithms,* edited by L. Eshelman, San Francisco: Morgan Kaufmann, pp. 73–80.

Kampfner, R. P., and M. Conrad (1983). "Computational Modeling of Evolutionary Learning Processes in the Brain," *Bull. Math. Biol.,* Vol. 45:6, pp. 931–968.

Kieras, D. E. (1976). "Automata and S-R Models," *J. of Mathematical Psychology,* Vol. 13, pp. 127–147.

Klopf, A. H. (1965). "Evolutionary Pattern Recognition Systems," Tech. Report on Contract AF-AFOSR-978-65, Univ. Illinois, Chicago, IL.

Klopf, A. H., and E. Gose (1969). "An Evolutionary Pattern Recognition Network," *IEEE Trans. Systems Science and Cybernetics,* Vol. SSC-5:3, pp. 247–250.

Koehler, T. L. (1958a). "Evolutionary Operation: Some Actual Examples," *Sec. Stevens Symp.,* pp. 5–8.

Koehler, T. L. (1958b). "Evolutionary Operation: A Program for Optimizing Plant Operation," *Trans. All-Day Rutgers Qual. Contr. Conf.,* American Soc. Qual. Cont., pp. 25–45.

Koza, J. R. (1989). "Hierarchical Genetic Algorithms Operating on Populations of Computer Programs," In *Proc. 11th Intern. Joint Conf. on Artificial Intelligence,* edited by N. S. Sridharan, San Mateo, CA: Morgan Kaufmann, pp. 768–774.

Koza, J. R. (1992). *Genetic Programming: On the Programming of Computers by Means of Natural Selection,* Cambridge, MA: MIT Press.

Kreinovich, V., C. Quintana, and O. Fuentes (1993). "Genetic Algorithms—What Fitness Scaling Is Optimal," *Cybernetics and Systems,* Vol. 24:1, pp. 9–26.

Kristinsson, K., and G. A. Dumont (1992). "System Identification and Control Using Genetic Algorithms," *IEEE Trans. Sys., Man and Cyber.,* Vol. 22:5, pp. 1033–1046.

Kuhn, W., and A. Visser (1992). "Identification of the System Parameter of a 6 Axis Robot with the Help of an Evolution Strategy," *Robotersysteme,* Vol. 8:3, pp. 123–133.

Lai, L. L. (1998). *Intelligent System Applications in Power Engineering: Evolutionary Programming and Neural Networks,* New York: John Wiley.

Lai, L. L., and J. T. Ma (1996). "Application of Evolutionary Programming to Transient and Subtransient Parameter Estimation," *IEEE Trans. Energy Conversion,* Vol. 11, pp. 523–530.

Lai, L. L., J. T. Ma, and K. P. Wong (1996). "Evolutionary Programming for Economic Dispatch of Units with Non-Smooth Input-Output Characteristic Functions," In *Proc. 12th Power Systems Computation Conference, PSCC,* pp. 492–498.

Lenat, D. B. (1983). "The Role of Heuristic in Learning by Discovery: Three Case Studies," In *Machine Learning,* edited by R. S. Michalski, J. G. Carbonell, and T. M. Mitchell, Palo Alto, CA: Tioga Publishing, pp. 243–306.

Liepens, G. E., M. R. Hilliard, M. Palmer, and G. Rangarajan (1991). "Credit Assignment and Discovery in Classifier Systems," *Intern. J. of Intelligent Sys.,* Vol. 6:1, pp. 55–69.

Liepens, G. E., and M. D. Vose (1991). "Deceptiveness and Genetic Algorithm Dynamics," In *Foundations of Genetic Algorithms,* edited by G.J.E. Rawlins, San Mateo, CA: Morgan Kaufmann, pp. 36–52.

Lindsay, R. K. (1968). "Artificial Evolution of Intelligence," *Contemp. Psychology,* Vol. 13:3, pp. 113–116.

Lohmann, R. (1992). "Structure Evolution and Incomplete Induction," In *Parallel Problem Solving from Nature 2,* edited by R. Männer and B. Manderick, Amsterdam: North-Holland, pp. 175–186.

Luke, S., and L. Spector (1998). "A Revised Comparison of Crossover and Mutation in Genetic Programming," In *Genetic Programming 98: Proceedings of the Third Annual Genetic Programming Conference,* edited by J. R. Koza, W. Banzhaf, K. Chellapilla, K. Deb, M. Dorigo, D. B. Fogel, M. H. Garzon, D. E. Goldberg, H. Iba, and R. L. Riolo, San Francisco: Morgan Kaufmann, pp. 208–213.

Lutter, B. E., and R. C. Huntsinger (1969). "Engineering Applications of Finite Automata," *Simulation,* Vol. 13, pp. 5–11.

Lyle, M. R. (1972). "An Investigation into Scoring Techniques in Evolutionary Programming," Master's Thesis, New Mexico State University, Las Cruces, NM.

Mallows, C. L. (1973). "Some Comments on C_p," *Technometrics,* Vol. 15, pp. 661–675.

Männer, R., and B. Manderick (eds.) (1992). *Parallel Problem Solving from Nature 2,* Amsterdam: North-Holland.

Matsumura, T., M. Nakamura, J. Okech, and K. Onaga (1998). "A Parallel and Distributed Genetic Algorithm on Loosely-Coupled Multiprocessor Systems," *IEICE Trans. Fund. Electron., Comm., and Comp. Sci.,* Vol. E81-A:4, pp. 540–546.

Mayoh, B. (1996). "Artificial Life and Pollution Control: Explorations of a Genetic Algorithm System on the Highly Parallel Connection Machine," In *Perspectives of System Informatics,* edited by D. Bjorner, M. Broy, and I. V. Pottosin, Berlin: Springer, pp. 68–79.

Mayr, E. (1963). *Animal Species and Evolution,* Cambridge, MA: Belknap Press.

McDonnell, J. R., B. D. Andersen, W. C. Page, and F. Pin (1992). "Mobile Manipulator Configuration Optimization Using Evolutionary Programming," In *Proc. of First Ann. Conf. on Evolutionary Programming,* edited by D. B. Fogel and W. Atmar, La Jolla, CA: Evolutionary Programming Society, pp. 52–62.

McDonnell, J. R., and W. C. Page (1990). "Mobile Robot Path Planning Using Evolutionary Programming," In *Proc. of the 24th Asilomar Conf. on Signals, Syst., and Computers,* edited by R. R. Chen, Pacific Grove, CA: Maple Press, pp. 1025–1029.

McDonnell, J. R., R. G. Reynolds, and D. B. Fogel (eds.) (1995). *Evolutionary Programming IV: Proceedings of the Fourth Annual Conference on Evolutionary Programming,* Cambridge, MA: MIT Press.

McDonnell, J. R., and D. Waagen (1994). "Evolving Recurrent Perceptrons for Time-Series Prediction," *IEEE Trans. Neural Networks,* Vol. 5:1, pp. 24–38.

Michalewicz, Z. (1992). *Genetic Algorithms + Data Structures = Evolution Programs,* New York: Springer-Verlag.

Michie, D. (1970). "Future for Integrated Cognitive Systems," *Nature,* Vol. 228, pp. 717–722.

Minsky, M. L. (1961). "Steps Toward Artificial Intelligence," *Proc. of the IRE,* Vol. 49:1, pp. 8–30.

Minsky, M. L., and O. G. Selfridge (1961). "Learning in Random Nets," In *Proc. 4th Lond. Symp. on Information Theory,* edited by C. Cherry, London: Butterworths.

Montana, D. J. (1991), "Automated Parameter Tuning for Interpretation of Synthetic Images," In *Handbook of Genetic Algorithms,* edited by L. Davis, New York: Van Nostrand Reinhold, pp. 282–311.

Montez, J. (1974). "Evolving Automata for Classifying Electrocardiograms," Master's Thesis, New Mexico State University, Las Cruces, NM.

Moosburger, R., C. Kostrzewa, G. Fischbeck, and K. Petermann (1997). "Shaping the Digital Optical Switch Using Evolution Strategies and BPM," *IEEE Phot. Tech. Lett.,* Vol. 9:11, pp. 1484–1486.

Mühlenbein, H., M. Schomisch, and J. Born (1991). "The Parallel Genetic Algorithm as Function Optimizer," *Parallel Computing,* Vol. 17, pp. 619–632.

Nelson, K. (1994). "Function Optimization and Parallel Evolutionary Programming on the MasPar MP-1," In *Proc. Third Ann. Conf. on Evol. Prog.,* edited by A. V. Sebald and L. J. Fogel, River Edge, NJ: World Scientific, pp. 324–334.

Ogawa, S., T. Watanabe, K. Yasuda, and R. Yokoyama (1997). "A Study for Keeping Generalization Ability of Multilayered Neural Networks Using Evolution Strategies," *Trans. Inst. Elect. Eng. of Japan, Part C,* Vol. 117-C:2, pp. 143–149.

Ostermeier, A. (1992). "An Evolution Strategy with Momentum Adaptation of the Random Number Distribution," In *Parallel Problem Solving from Nature 2,* edited by R. Männer and B. Manderick, Amsterdam: North-Holland, pp. 197–206.

Page, W. C., B. D. Andersen, and J. R. McDonnell (1992). "An Evolutionary Programming Approach to Multi–Dimensional Path Planning," In *Proc. of First Ann. Conf. on Evolutionary Programming,* edited by D. B. Fogel and W. Atmar, La Jolla, CA: Evolutionary Programming Society, pp. 63–70.

Pincus, M. (1970). "An Evolutionary Strategy," *J. Theoret. Biol.,* Vol. 28, pp. 483–488.

Pitney, G., T. R. Smith, and D. Greenwood (1990). "Genetic Design of Processing Elements for Path Planning Networks," In *Proc. of the Intern. Joint Conf. on Neural Networks* 1990, Vol. III, Piscataway, NJ: IEEE Press, pp. 925–932.

Porto, V. W., and L. J. Fogel (1997). "Evolution of Intelligently Interactive Behaviors for Simulated Forces," In *Evolutionary Programming VI,* edited by P. J. Angeline, R. G. Reynolds, J. R. McDonnell, and R. Eberhart, Berlin: Springer, pp. 419–429.

Porto, V. W., N. Saravanan, D. Waagen, and A. E. Eiben (eds.) (1998). *Evolutionary Programming VII,* Berlin: Springer.

Qi, X., and F. Palmieri (1993). "Adaptive Mutation in the Genetic Algorithm," In *Proc. of the Sec. Ann. Conf. on Evolutionary Programming,* edited by D. B. Fogel and W. Atmar, La Jolla, CA: Evolutionary Programming Society, pp. 192–196.

Rada, R. (1981). "Evolution and Gradualness," *BioSystems,* Vol. 14, pp. 211–218.

Ray, T. S. (1991a). "An Approach to the Synthesis of Life," In *Artificial Life II,* edited by C. G. Langton, C. Taylor, J. D. Farmer, and S. Rasmussen, Reading, MA: Addison-Wesley, pp. 371–408.

Ray, T. S. (1991b). "Is It Alive Or Is It GA?" In *Proceedings of the Fourth International Conference on Genetic Algorithms,* edited by R. K. Belew and L. B. Booker, San Mateo, CA: Morgan Kaufmann, pp. 527–534.

Ray, T. S. (1994). Personal Communication, ATR, Kyoto, Japan.

Rechenberg, I. (1965). "Cybernetic Solution Path of an Experimental Problem," Royal Aircraft Establishment, Library Translation No. 1122, August.

Rechenberg, I. (1973). *Evolutionsstrategie: Optimierung Technisher Systeme nach Prinzipien der Biologischen Evolution,* Fromman-Holzboog Verlag, Stuttgart.

Rechenberg, I. (1993). Personal Communication Technical University of Berlin, Germany.

Reed, J., R. Toombs, and N. A. Barricelli (1967). "Simulation of Biological Evolution and Machine Learning," *J. Theoretical Biology,* Vol. 17, pp. 319–342.

Richter, R., and W. Hofmann (1997). "Evolution Strategies Applied to Controls on a Two Axis Robot," In *Computational Intelligence Theory and Applications: International Conference, 5th Fuzzy Days,* edited by B. Reusch, Berlin: Springer, pp. 434–443.

Riolo, R. (1991). "Modeling Simple Human Category Learning with a Classifier System," In *Proc. of the Fourth Intern. Conf. on Genetic Algorithms,* edited by R. K. Belew and L. B. Booker, San Mateo, CA: Morgan Kaufmann, pp. 324–333.

Rissanen, J. (1978). "Modeling by Shortest Data Description," *Automatica,* Vol. 14, pp. 465–471.

Rizki, M. M., and M. Conrad (1985). "Evolve III: A Discrete Events Model of an Evolutionary Ecosystem," *BioSystems,* Vol. 18, pp. 121–133.

Root, R. (1970). "An Investigation of Evolutionary Programming," Master's Thesis, New Mexico State University, Las Cruces, NM.

Rosenberg, R. (1967). "Simulation of Genetic Populations with Biochemical Properties," Doctoral Dissertation, Univ. Michigan, Ann Arbor.

Rudolph, G. (1994). "Convergence Properties of Canonical Genetic Algorithms," *IEEE Trans. on Neural Networks,* Vol. 5:1, pp. 96–101.

Rudolph, G. (1997). *Convergence Properties of Evolutionary Algorithms,* Hamburg, Germany: Verlag Dr. Kovac.

Salomon, R. (1996). "Re-Evaluating Genetic Algorithm Performance under Coordinate Rotation of Benchmark Functions. A Survey of Some Theoretical and Practical Aspects of Genetic Algorithms," *BioSystems,* Vol. 39:3, pp. 263–278.

Saravanan, N., and D. B. Fogel (1994). "Learning of Strategy Parameters in Evolutionary Programming: An Empirical Study," In *Proc. of the Third Ann. Conf. on*

Evolutionary Programming, edited by A. V. Sebald and L. J. Fogel, River Edge, NJ: World Scientific, pp. 269–280.

Saravanan, N., D. B. Fogel, and K. M. Nelson (1995). "A Comparison of Methods for Self-Adaptation in Evolutionary Algorithms," *BioSystems,* Vol. 36, pp. 157–166.

Schaffer, J. D., and L. J. Eshelman (1991). "On Crossover as an Evolutionarily Viable Strategy," In *Proc. of the Fourth Intern. Conf. on Genetic Algorithms,* edited by R. K. Belew and L. B. Booker, San Mateo, CA: Morgan Kaufmann, pp. 61–68.

Schraudolph, N. N. (1993). Personal Communication, Salk Institute, La Jolla, CA.

Schraudolph, N. N., and R. K. Belew (1992). "Dynamic Parameter Encoding for Genetic Algorithms," *Machine Learning,* Vol. 9:1, pp. 9–21.

Schwefel, H.-P. (1965). "Kybernetische Evolution als Strategie der Experimentellen Forschung in der Strömungstechnik," Diploma Thesis, Technical University of Berlin.

Schwefel, H.-P. (1981). *Numerical Optimization of Computer Models,* Chichester, U.K.: John Wiley.

Schwefel, H.-P., and G. Rudolph (1995). "Contemporary Evolution Strategies," In *Advances in Artificial Life. Third European Conference on Artificial Life Proceedings,* edited by F. Moran, A. Moreno, J. J. Merelo, and P. Chacon, Berlin: Springer, pp. 893–907.

Sebald, A. V., and K. Chellapilla (1998). "On Making Problems Evolutionary Friendly. Part 1: Evolving the Most Convenient Representation," In *Evolutionary Programming VII,* edited by V. W. Porto, N. Saravanan, D. Waagen, and A. E. Eiben, Berlin: Springer, pp. 271–280.

Sebald, A. V., and L. J. Fogel (eds.) (1994). *Proceedings of the Third Annual Conference on Evolutionary Programming,* River Edge, NJ: World Scientific Publishing.

Sebald, A. V., and J. Schlenzig (1994). "Minimax Design of Neural Net Controllers for Highly Uncertain Plants," *IEEE Trans. Neural Networks,* Vol. 5:1, pp. 73–82.

Shapiro, B. A., and J. C. Wu (1996). "An Annealing Mutation Operator in the Genetic Algorithms for RNA Folding," *Comp. Applic. Biosciences,* Vol. 12:3, pp. 171–180.

Solis, F. J., and R.J.-B. Wets (1981). "Minimization by Random Search Techniques," *Math. Operations Research,* Vol. 6, pp. 19–30.

Solomonoff, R. J. (1966). "Some Recent Work in Artificial Intelligence," *Proc. of the IEEE,* Vol. 54:12, pp. 1687–1697.

Spears, W. M., and K. A. De Jong (1991). "On the Virtues of Parameterized Uniform Crossover," In *Proc. of the Fourth Intern. Conf. on Genetic Algorithms,* edited by R. K. Belew and L. B. Booker, San Mateo, CA: Morgan Kaufmann, pp. 230–236.

Spiessens, P., and B. Manderick (1991). "A Massively Parallel Genetic Algorithm: Implementation and First Analysis," In *Proc. of the Fourth Inter. Conf. on Genetic Algorithms,* edited by R. K. Belew and L. B. Booker, San Mateo, CA: Morgan Kaufmann, pp. 279–286.

Syswerda, G. (1989). "Uniform Crossover in Genetic Algorithms," In *Proc. of the Third Int. Conf. on Genetic Algorithms,* edited by J. D. Schaffer, San Mateo, CA: Morgan Kaufmann, pp. 2–9.

Syswerda, G. (1991). "Schedule Optimization Using Genetic Algorithms," In *Handbook of Genetic Algorithms,* edited by L. Davis, New York: Van Nostrand Reinhold, pp. 332–349.

Takeuchi, A. (1980). "Evolutionary Automata—Comparison of Automaton Behavior and Restle's Learning Model," *Information Sciences,* Vol. 20, pp. 91–99.

Trellue, R. E. (1973). "The Recognition of Handprinted Characters Through Evolutionary Programming," Master's Thesis, New Mexico State University, Las Cruces, NM.

Turing, A. M. (1950). "Computing Machinery and Intelligence," *Mind,* Vol. 59, pp. 433–460.

Vignaux, G. A., and Z. Michalewicz (1991). "A Genetic Algorithm for the Linear Transportation Problem," *IEEE Trans. on Sys., Man and Cybern.,* Vol. 21:2, pp. 445–452.

Vincent, R. W. (1976). "Evolving Automata Used for Recognition of Digitized Strings," Master's Thesis, New Mexico State University, Las Cruces, NM.

Voigt, H.-M., W. Ebeling, I. Rechenberg, and H.-P. Schwefel (eds.) (1996). *Parallel Problem Solving from Nature IV,* Berlin: Springer.

Whitley, L. D. (1991). "Fundamental Principles of Deception in Genetic Search," In *Foundations of Genetic Algorithms,* edited by G.J.E. Rawlins, San Mateo, CA: Morgan Kaufmann, pp. 221–241.

Whitley, L. D., and M. D. Vose (eds.) (1995). *Foundations of Genetic Algorithms 3,* San Mateo, CA: Morgan Kaufmann.

Wilde, D. J. (1964). *Optimum Seeking Methods,* Englewood Cliffs, NJ: Prentice-Hall.

Wilde, D. J., and C. S. Beightler (1967). *Foundations of Optimization,* Englewood Cliffs, NJ: Prentice-Hall.

Williams, G. L. (1977). "Recognition of Hand–Printed Numerals Using Evolving Automata," Master's Thesis, New Mexico State University, Las Cruces, NM.

Wilson, S. W. (1995). "Classifier Fitness Based on Accuracy," *Evolutionary Computation,* Vol. 3:2, pp. 149–175.

Wong, K. P., and J. Yuryevich (1998). "Evolutionary Programming Based Algorithm for Environmentally Constrained Economic Dispatch," *IEEE Trans. Power Systems,* Vol. 13, pp. 301–306.

Wright, A. H. (1991). "Genetic Algorithms for Real Parameter Optimization," In *Foundations of Genetic Algorithms,* edited by G.J.E. Rawlins, San Mateo, CA: Morgan Kaufmann, pp. 205–218.

Yang, H.-T., P.-C. Yang, and C.-L. Huang (1996). "Evolutionary Programming Based Economic Dispatch for Units with Non-Smooth Fuel Cost Functions," *IEEE Trans. Power Systems,* Vol. 11, pp. 112–118.

Yao, X., and Y. Liu (1997). "A New Evolutionary System for Evolving Artificial Neural Networks," *IEEE Transactions on Neural Networks,* Vol. 8:3, pp. 694–713.

Zacharias, C. R., M. R. Lemes, and A. D. Pino (1998). "Combining Genetic Algorithm and Simulated Annealing: A Molecular Geometry Optimization Study," *Theochem: J. Mol. Structure,* Vol. 430, pp. 29–39.

4

THEORETICAL AND EMPIRICAL PROPERTIES OF EVOLUTIONARY COMPUTATION

4.1 THE DIFFICULTIES

Evolutionary methods of optimization, indeed all optimization algorithms, can be analyzed and evaluated. Inquiries into the efficiency of the procedures, including the rate of convergence, the quality of the evolved solution, the computational requirements, and other properties, can be focused along two lines of investigation: theoretical and empirical. The theoretical approach attempts to discover mathematical truths about the algorithms that will hold in a reasonably broad domain of applications. The empirical approach attempts to assay the performance of each instance of an evolutionary algorithm in specific domains through statistical means. Both procedures are inherently limited.

Mathematical proofs regarding the properties of algorithms are always more powerful than mere empirical evidence. But it may be very difficult or even impossible to formulate a purely theoretical analysis. Evolutionary algorithms generally incorporate complex nonlinear stochastic processes. To make analysis tractable, the actual procedure is often simplified, but then the true algorithm is not being studied. One may legitimately question the degree to which what may be proved for the simplified procedure will hold for the true procedure.

Similarly, through statistical analysis, the algorithm can be tested over successive trials on a specific problem, and a probabilistic description of the procedure's performance can be determined. This method can be useful for comparing the efficiency of two or more algorithms applied to the same task. But again, the performance of an algorithm on one sample problem may not immediately convey information on how it will perform on any other problem. Indeed, this can be proved mathematically (Wolpert and Macready, 1997): Across all problems, all algorithms that do not resample points in a search space perform identically on average.

In the extreme, it may be argued that if a theoretical analysis is conducted on a procedure that is even the slightest bit a simplification of the real process, then nothing is truly learned about the real process. Furthermore, one may argue that empirical evidence is useless for assessing the general performance of an algorithm because it is strictly limited to the domain of experimentation and may not be more broadly applicable.

The position adopted here will be an intermediate one. It is possible to gain insight into the behavior of a complex algorithm by proving mathematical properties associated with simplifications of the algorithm. Furthermore, it is possible to generalize about the properties of a complex algorithm by assessing its performance across a variety of domains. Of course, such generalization may be in error. Yet it is only through this two-pronged investigation that real insight into the effectiveness and efficiency of evolutionary computation can be garnered. To forsake one approach in favor of the other would be foolish.

4.2   THEORETICAL ANALYSIS OF EVOLUTIONARY COMPUTATION

4.2.1   The Framework for Analysis

There are many variants of evolutionary computation. In order to begin with the most high-level analysis of these procedures before turning our attention to more specific nuances of particular implementations, consider an evolutionary algorithm that operates on a population of contending solutions to a problem. The objective is to find the vector $\mathbf{x}$ that minimizes or maximizes a functional $f(\mathbf{x})$. Each extant solution $\mathbf{x}$ is measured in quality by its corresponding value $f(\mathbf{x})$. Some selection rule is applied to determine which vectors in the population will serve as parents based on their respective quality measures. Variation operators are then applied to generate new solutions, and the process iterates over successive generations.

This framework encompasses the majority of current implementations of evolutionary computation. Each vector $\mathbf{x}$ might be a collection of real values, binary digits, a mixture of these, or something more complicated representing a neural network, finite-state machine, or symbolic expression. Selection might come in the form of truncating out some percentage of the worst-scoring solutions, or reproducing each existing solution some multiple of times based on its quality relative to the mean quality of all solutions in the population, or using several other methods. New solutions might be generated by making random modifications to a single "parent" or by blending or crossing between two or more parents, or other even more sophisticated techniques (e.g., majority logic). Specifically, this framework includes the canonical algorithms offered by Fraser (1957a), Bremermann (1958, 1962), Fogel (1964), Rechenberg (1965), Reed et al. (1967), Holland (1975), and many modern variations of these and other early procedures.

4.2.2 Convergence in the Limit

The efficacy of evolutionary algorithms can be assessed in a variety of ways. One obvious criterion relates to convergence. Holland (1975, pp. 16–17, 124) suggested that convergence was not a useful performance measure. He noted that there are procedures for searching large finite spaces that converge to global optima but require excessively long periods of time (e.g., enumeration). Such procedures are generally useless in complex problems. But any optimization algorithm that avoids entrapment in local optima possesses an infinity-to-one speed advantage over a search algorithm that becomes easily entrapped (Atmar, 1990). Holland (1975, p. 104; and others) proposed that certain variations of evolutionary algorithms do not get entrapped in local optima when further improvements are possible. If this is correct, then convergence should be a useful measure of performance and a reasonable measure for assessing the efficiency of these approaches in searching complex state spaces, as well as for making comparisons to other more classic methods.

The most common approach to assessing the long-term behavior of an evolutionary algorithm relies on the use of a Markov chain. This is a particular type of stochastic process that is time invariant and memoryless. For every possible "state" of the process, the transition probabilities from state to state do not change over time, and these probabilities are only dependent on the current state and not on any prior state that was visited previously. Many versions of evolutionary algorithms can be placed in the form of a Markov chain, thereby providing a means of illuminating some important results concerning their limiting behavior.

To illustrate, consider the following specific procedure:

1. Initialize a population of solutions. Each individual is encoded in an alphabet of specified cardinality and describes all the necessary variables to implement a candidate solution.

2. Evaluate each solution in the population.

3. Create new solutions by varying current solutions.

4. Delete members of the population to make room for the new solutions.

5. Evaluate the new solutions and insert them into the population.

6. If the available time has expired, halt and return the best solution; if not, go to 3.

Let us first consider the case where solutions are encoded as binary digits.

The number of bits that must be used to describe each variable parameter is problem dependent. Let each solution in the population of m such solutions, x_i, $i = 1, \ldots, m$, be a string of symbols $\{0, 1\}$ of length k. Often, the initial population of m solutions is selected completely at random, with each bit of each solution having a 50 percent chance of taking the symbol 0.

Evaluation is made with respect to some extrinsically defined payoff (cost) function. This function can itself be a function of the evolving population, although this procedure is not practiced traditionally. For the sake of discussion, consider the function to be a real-valued mapping (i.e., a functional) from the bit strings of length k to $\Re$. For example, the payoff function, $f(\mathbf{y})$, might be:

$$f(\mathbf{y}) = y_1^2 + y_2^2$$

where y_1 is determined from the first j bits of solution x_i and y_2 is determined from the remaining bits. There are no restrictions on which bits must be used for each parameter, but once chosen they should remain fixed for the duration of the simulation.

Consider the cases where new solutions are created by using "reproduction with emphasis" applied to the existing parents. There are many ways to accomplish this, but one common method is termed *roulette wheel parent selection* (see Chapter 3 and also Davis, 1991, p. 13). This method proceeds by summing the fitnesses of all the population members. Call this value S. A uniformly distributed random number u ranging over $[0, S]$ is selected. Reproduction is subsequently applied to the first population member whose fitness, summed with the fitnesses of the preceding population members, is greater than or equal to u. Therefore, the chance of a parent being selected for reproduction is directly proportional to its fitness (which is constrained to be positive). Reproduction is carried out until the population size of the next generation is filled.

The resulting population of strings is then modified according to variation operators. Let us first consider the case where these are limited to bit mutations and one-point crossover. Other forms of crossover, such as two-point or uniform crossover, have been proposed but will not be analyzed from a theoretical perspective here. Experiments with these other operators are detailed later in this chapter.

Bit mutation is accomplished through one of two methods: by considering each bit of each solution in the population and then with a certain probability either (1) flipping it to the complementary value (i.e., changing the bit) or (2) randomly choosing a value with a certain probability (e.g., 50 percent chance of changing the bit). Within this framework, the probability of bit mutation is often set at a very small number (on the order of 10^{-2} to 10^{-3}, Davis, 1991, p. 15; following earlier suggestions from Holland, 1975, pp. 109–111, that mu-

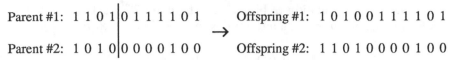

Crossover Point

Parent #1: 1 1 0 1 | 0 1 1 1 1 0 1   Offspring #1: 1 0 1 0 0 1 1 1 1 0 1

→

Parent #2: 1 0 1 0 | 0 0 0 0 1 0 0   Offspring #2: 1 1 0 1 0 0 0 0 1 0 0

Figure 4-1 The one-point crossover operator is applied to two parents and creates two offspring. A crossover point is selected, and the sections of the parental strings that precede the crossover point are exchanged.

tation was only a background operator and not of primary importance). Both methods of implementation can be made equivalent by adjusting the mutation rate. Mutation is employed in order to ensure that all bits have some probability of entering the population.

One-point crossover creates two offspring from two parents. The parents are selected at random (typically in accordance with a uniform distribution[1]), a crossover point, c_x, is selected at random (also in accordance with a uniform distribution), and two offspring are made by concatenating the bits that precede c_x in the first parent with those that follow (and include) c_x in the second parent, and also performing the obverse operation (Figure 4-1).

Solutions are deleted from the population in order to make room for the new solutions generated through reproduction, mutation, and crossover. This process is then iterated until the known global optimum solution is discovered or until the available computer time is exhausted.

Consider a state space of possible solutions to be coded as strings of k bits $\{0, 1\}$. Let there be a population of m such bit strings, and let each possible configuration have a fitness f_i, $i = 1, \ldots, 2^k$. Let f^* be the globally optimum value. To simplify this first analysis, let an evolutionary algorithm with only one-point crossover (no bit mutation or other search operator) and selection (differential reproduction) operate on the bit strings.

This search can be formulated as a finite-dimension Markov chain (Fogel, 1992, 1994a; others, including Eiben et al., 1991, Davis and Principe, 1993, and Rudolph, 1994a, have also studied evolutionary algorithms within the framework of Markov chains—see Goodman, 1988, for a discussion of Markov chains) characterized by a state vector π and a transition matrix P. Given π, a row vector describing the probability of being in each state, πP yields the probability of being in each state after one transition. Thus, the probability of being in each state after n transitions is πP^n.

---

[1] Rather than apply crossover after reproduction with emphasis has generated a completely new population, an essentially equivalent procedure could be implemented to create offspring immediately after reproducing each pair of parents.

Here, the states of the chain can be defined by every possible configuration of an entire population of bit strings. There are 2^{mk} such states. For example, if $k = 2$ and $m = 2$, then the following possibilities exist:

$$(00, 00), (00, 01), (00, 10), (00, 11),$$
$$(01, 00), (01, 01), (01, 10), (01, 11),$$
$$(10, 00), (10, 01), (10, 10), (10, 11),$$
$$(11, 00), (11, 01), (11, 10), (11, 11).$$

Each of these 16 collections of pairs of bit strings represents a possible state in the chain. For practical problems, often $m = 100$ and $k > 50$; thus, there may be more than 2^{5000} possible states in the chain, yet the chain is of finite dimension and possesses the *time homogeneous* and *no memory* properties required (Goodman, 1988, p. 127).

A Markov chain is said to be *irreducible* if every state communicates with every other state. Two states, i and j, are said to *communicate* if there is at least one path from i to j and vice versa. Such a path may require multiple steps (i.e., there exists $P_{ij}^x > 0$, $P_{ji}^y > 0$, for some $0 \le x < M$, $0 \le y < M$, where there are M states and P_{ij} is the probability of transitioning from state i to state j in one step).

If the above evolutionary algorithm operates with only crossover and selection, then the chain defined by that algorithm is not irreducible as there are absorbing states (i.e., states that do not communicate with any other state). Let the only *ergodic* classes (i.e., sets of states from which every path leads to a state in the set) be single absorbing states; no oscillations between states will be considered.

A state in this chain will be absorbing if all members of the population are identical. Under such conditions, crossing over bit strings in the population simply yields the original population. Otherwise, a state is transient. As time progresses, the behavior of the chain will be described by (1) a transition to an absorbing state, (2) a transition to a state from which there may be a transition to an absorbing state with some nonzero probability, or (3) a transition to a state from which there is no probability of transitioning to an absorbing state in a single step. Thus, the states can be indexed such that the state transition matrix, P, for the chain satisfies:

$$P = \begin{bmatrix} I_a & 0 \\ R & Q \end{bmatrix} \tag{4-1}$$

where I_a is an $a \times a$ identity matrix describing its absorbing states, R is a $t \times a$ transition submatrix describing transitions to an absorbing state, Q is a $t \times t$ transition submatrix describing transitions to transient states and not to an absorbing state, and a and t are positive integers.

The behavior of such a chain satisfies:

$$P^n = \begin{bmatrix} I_a & 0 \\ N_n R & Q^n \end{bmatrix} \tag{4-2}$$

where P^n is the n-step transition matrix, $N_n = I_t + Q + Q^2 + Q^3 + \ldots + Q^{n-1}$, and I_t is the $t \times t$ identity matrix. As n tends to infinity:

$$\lim_{n \to \infty} P^n = \begin{bmatrix} I_a & 0 \\ (I_t - Q)^{-1}R & 0 \end{bmatrix} \tag{4-3}$$

(Goodman, 1988, p. 158). The matrix $(I_t - Q)^{-1}$ is guaranteed to exist (Goodman, 1988, p. 160). Therefore, given infinite time, the chain will transition with probability one to an absorbing state. Note that there is a nonzero probability that the absorbing state may not be the global best state unless all the absorbing states are globally optimal (every homogeneous population is an absorbing state).

The results may be summarized by the following:

THEOREM
Let $\Gamma \equiv \{0, 1\}$, m be the number of solutions in Γ^k maintained at each iteration, the loss function $L : (\Gamma^k)^m \to \Re_d$, where $\Re_d$ describes the set of real numbers representable in a given digital machine, and $L(\gamma), \gamma \in (\Gamma^k)^m$, be single-valued. After n iterations, the evolutionary algorithm relying only on (one-point) crossover and roulette wheel selection and without bit mutation (see above) arrives at a state γ:

$$\gamma \in (\Gamma^k)^m \ni \Pr(\gamma \in A) = \sum_{i=1}^{a} (\pi * P^n)_i \tag{4-4}$$

$$= \sum_{i=1}^{a} \left(\pi * \begin{bmatrix} I_a \\ N_n R \end{bmatrix} \right)_i \tag{4-5}$$

where $(\pi * P^n)_i$ denotes the ith element in the row vector $(\pi * P^n)$, A is the set of all absorbing states, $\pi*$ is the row vector describing the initial probability of being in each state, P is the transition matrix defining the Markov chain of the form:

$$P = \begin{bmatrix} I_a & 0 \\ R & Q \end{bmatrix} \tag{4-6}$$

$N_n = I_t + Q + Q^2 + Q^3 + \ldots + Q^{n-1}$, and I_a is an $a \times a$ identity matrix. The limit of the probability of absorption is

$$\lim_{n \to \infty} \sum_{i=1}^{a} \left(\pi * \begin{bmatrix} I_a \\ N_n R \end{bmatrix} \right)_i = \sum_{i=1}^{a} \left(\pi * \begin{bmatrix} I_a \\ (I - Q)^{-1}R \end{bmatrix} \right)_i = 1 \tag{4-7}$$

It is natural to examine the number of absorbing states in such a chain. Absorbing states are those in which each bit string in the population is identical. There are 2^k such absorbing states. The total number of states is 2^{mk}, and this leads to two observations: The density of absorbing states decreases exponentially with the length of the bit string (i.e., $2^{k(1-m)}$), but the actual number of absorbing states increases exponentially with the length of the bit string (i.e., 2^k). For codings other than bit strings, the base of 2 will change but the relationships will not.[2]

As the chain must eventually transition to an absorbing state under the rules of the transition matrix described above (Eq. 4-3), if the transitions between states are of equal probability, the time required to find an absorbing state may increase exponentially with k. When such a state is found, the likelihood of it being a globally optimum state may decrease exponentially with k. Of course, for matrices in which the transition probabilities are not equal, as would be the case in practice, the specific entries in each matrix will determine the mean waiting time before absorption and the likelihood of discovering a global optimum versus becoming trapped at a suboptimum. Furthermore, should any particular bit be fixed across the population to a value that is not associated with a global optimum solution, a global optimum will never be discovered.

If each bit is given a probability of mutation (i.e., flipping), $p_m \in (0, 1]$, then the absorbing states of the transition matrix defined by the evolutionary algorithm operating only with crossover will become transient. Because of roulette wheel selection, any solution in the population may be lost at any generation, and bit mutation offers the possibility of generating any solution from the available search space in a single step. All globally optimal states can be collected and described as a single state. The mean time before entering this state if the chain starts in state i can be calculated by

$$\sum_{j \in T} n_{ij} \qquad (4\text{-}8)$$

where $N = (I - Q)^{-1} = [n_{ij}]$, and T is the set of all transient states (Goodman, 1988, p. 161). Thus, the specific number of steps taken until reaching a globally optimal state is highly dependent on the characteristics of P.

If the probability for bit mutation is set very low, this operation may indeed be viewed as a background operator, following Holland (1975, p. 111). It will be likely that the chain of states generated by high crossover rates and low probabilities of bit mutation will generally follow the chain generated by crossover alone, with mutation only serving to prevent complete stagnation (Holland, 1975, p. 111). Thus, the path that the evolutionary algorithm takes is likely to transition to a *meta-stable state* (an absorbing state of the chain defined by the evolutionary algorithm relying only on crossover), whereupon it will wait for mutation to affect the appropriate changes of bits in the population and then proceed to another such meta-stable state, and so forth (see Figure 4-2).

---

[2] The order of the elements in the population is generally irrelevant, and the procedure could perhaps be better characterized by describing the states in the chain as unordered multisets. The basic asymptotic convergence properties will remain unchanged under such a description, but the density of absorbing states will increase.

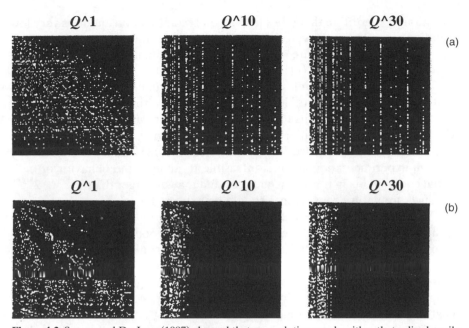

Figure 4-2 Spears and De Jong (1997) showed that an evolutionary algorithm that relies heavily on recombination is likely to stagnate in a state where the population is homogeneous. (a) The three graphs represent the Markov state transition at the 1st, 10th, and 30th generations when states are organized (from top-bottom and left-right) in terms of their mean fitness taken across all individuals in a population. The striated lines indicate limiting probabilities of being observed in each respective state. (b) The three graphs represent the same likelihoods, but the states have been reorganized in increasing order of average Hamming distance between individuals in the populations. The result indicates that most of the long-term probability mass is associated with states that are homogeneous. Spears and De Jong (1997) also noted that the amount of probability mass decreases with decreasing fitness of these homogeneous states. That is, the algorithm was most likely to stagnate at the homogeneous state with the global solution, was second most likely to stagnate at the homogeneous state with the second-best solution, and so forth.

Such transitions may require a very long waiting time, depending on the number of bits that must be flipped simultaneously. The probability of flipping b specific bits in a single solution (i.e., a bit string) and not flipping any others is $p_m^b (1 - p_m)^{k-b}$, where p_m is the probability of flipping a single bit. As the length of the codings, k, grows longer, the number of meta-stable states increases exponentially and very long waiting times become virtually certain.

These meta-stable states are associated with the term *premature convergence*. When reaching any of these states, the algorithm has prematurely stalled at a solution that is not globally optimal and may not even be locally optimal (cf. Holland, 1975, p. 140). This is a common occurrence when using an evolutionary algorithm that relies heavily on recombination and uses very low mutation rates. The most typical means for circumventing this situation are either to restart the procedure or to employ some heuristics for continuing the search (e.g., hill climbing) (Davis, 1991, p. 26).

The specific waiting times in the meta-stable states are problem dependent. There may well be problems for which such waiting times are relatively short or for which the paths to meta-stable states are unlikely. The degree to which immensely long waiting times are endemic to evolutionary algorithms that place a strong reliance on crossover can be reasonably assessed only through experimentation, for it is infeasible to analyze the behavior of individual components in transition matrices that may be larger than $2^{5000} \times 2^{5000}$. But such premature convergence should be expected.

Note that the term *premature convergence* may be somewhat misleading in this context. Under roulette wheel selection, the above evolutionary algorithm is strictly divergent (Rudolph, 1994a): It does not converge at all, regardless of the initialization, crossover operator, and objective function. It does not generate a sequence of solutions that converges to any point in the sample space, including any global optima. Premature convergence merely indicates that the search becomes stagnant for a long and random amount of time. But the search can easily be made to globally converge by adding *elitist selection,* that is, incorporating a heuristic to always maintain the best solution in the population into successive generations. (This occurs naturally in forms of truncation and tournament selection; see Bäck, 1996.)

Presuming the use of elitist selection, form an equivalence class of all states that contain a global best solution and describe this class as a single state. For example, if the global optimum is uniquely $(0, 0)$, then all collections of vectors containing $(0, 0)$ are characterized as the same state. The Markov chain may then be written in the form

$$P = \begin{bmatrix} 1 & 0 \\ R & Q \end{bmatrix} \tag{4-9}$$

where P is the transition matrix, 1 is a 1×1 identity matrix describing the absorbing state, R is a strictly positive (all entries are greater than zero) $t \times 1$ transition submatrix, Q is a $t \times t$ transition submatrix, and t is a positive integer. The state containing global optima is the only absorbing state, and all other states are transient.

Asymptotic global convergence of the evolutionary algorithm is then transparent as every absorbing chain will reach an absorbing state; in this case, there is just one such state. Furthermore, the matrix P is a special case of the matrix in Eq. 4-1:

$$P = \begin{bmatrix} I_a & 0 \\ R & Q \end{bmatrix} \tag{4-10}$$

It has already been shown (Eq. 4-2) that a Markov chain described by P will proceed as

$$P^n = \begin{bmatrix} I_a & 0 \\ N_n R & Q^n \end{bmatrix} \tag{4-11}$$

Thus (as in Eq. 4-3),

$$\lim_{n \to \infty} P^n = \begin{bmatrix} I_a & 0 \\ (I_t - Q)^{-1} R & 0 \end{bmatrix} \tag{4-12}$$

Therefore, the probability of absorption (that is, the probability of discovering a global optimum solution) for the chain with transition matrix P (Eq. 4-9) increases over n steps as a geometric series $(N_n R)$ converging to 1.0. This result can be summarized as follows:

THEOREM

Let $\Gamma \equiv \mathfrak{R}_d$, the set of elements of $\mathfrak{R}$ representable in a given digital machine, m, be the number of maintained solutions, $L \cdot (\Gamma^k)^m \to \Gamma$ be the loss function, and $L(\gamma), \gamma \in (\Gamma^k)^m$, be single-valued. Then, after n iterations, an evolutionary algorithm with elitism and complete communication between states arrives at a state:

$$\gamma \in (\Gamma^k)^m \ni \Pr(\gamma \in A) = \left(\pi^* P^n\right)_1 = \pi^* \begin{bmatrix} 1 \\ N_n R \end{bmatrix} \tag{4-13}$$

where the subscript 1 denotes the first entry in the row vector $(\pi^* P^n)$, A is the singleton set containing the absorbing state, π^* is the $1 \times (t + 1)$ row vector describing the initial probability of being in each state, P is the transition matrix defining the Markov chain, of the form

$$P = \begin{bmatrix} 1 & 0 \\ R & Q \end{bmatrix} \tag{4-14}$$

and $N_n = I_t + Q + Q^2 + Q^3 + \ldots + Q^{n-1}$.
Furthermore, the limit of the probability of absorption is

$$\lim_{n \to \infty} \pi^* \begin{bmatrix} 1 \\ N_n R \end{bmatrix} = \pi^* \begin{bmatrix} 1 \\ \vdots \\ 1 \end{bmatrix} = \sum_{i=1}^{t+1} \pi^*_i = 1 \tag{4-15}$$

This theorem directly indicates a convergence in probability, but it can be extended to imply convergence with probability one. Convergence in probability sufficiently fast implies convergence with probability one (Serfling, 1980, p. 10). More specifically, given a sequence x_n, if

$$\sum_{n=1}^{\infty} \Pr\left(|x_n - x| > \varepsilon\right) < \infty \text{ for every } \varepsilon > 0 \tag{4-16}$$

then x_n converges with probability one to x $(x_n \to x)$.

This result can be applied to obtain the following:

THEOREM

Given a $(t + 1) \times (t + 1)$ state transition matrix of the form of Eq. 4-9, where there is a single absorbing state and t transient states, and an initial $1 \times (t + 1)$ state probability row vector π^* describing a Markov chain, let π^*_t be the $1 \times t$ row vector describing the probability of starting in each of the transient states. Form the sequence

$$x_n(\omega) = \begin{cases} 1, \text{ if the chain has reached the absorbing} \\ \quad \text{state by the } n\text{th iteration;} \\ 0, \text{ otherwise} \end{cases}$$

The sequence $x_n(\omega)$ converges with probability 1 to 1.0.

Proof

Consider the two cases (1) $n = 0$ or $n = 1$, and (2) $n \geq 2$.

1. The probability of arriving at the absorbing state by the zeroth or first iteration is simply $\pi^* \begin{bmatrix} 1 \\ R \end{bmatrix} = p_a$.

2. To analyze the sequence for $n \geq 2$, it is useful to note the following. For $\mathbf{v}Q^m \neq 0$, where Q is a $t \times t$ substochastic matrix (Eq. 4-9), $\mathbf{v}$ is a $1 \times t$ row vector, $\mathbf{v} \neq 0$, $v_i \geq 0$ for all $i = 1, \ldots, t$, for all $m = 0, 1, 2, \ldots$ ($\mathbf{v}Q^m = 0$ indicates $x_n(\omega)$ converges with probability 1 to 1.0 trivially), since $q_{ij} < 1$,

$$\left\| \frac{\mathbf{v}}{\|\mathbf{v}\|_1} \cdot Q \right\|_2 < \left\| \frac{\mathbf{v}}{\|\mathbf{v}\|_1} \cdot Q \right\|_1 \tag{4-17}$$

$$= \left\| \pi Q \right\|_1, \pi = \frac{\mathbf{v}}{\|\mathbf{v}\|_1} \tag{4-18}$$

$$= \sum_{i=1}^{t} \pi_i \sum_{j=1}^{t} q_{ij} \tag{4-19}$$

$$\leq \sum_{i=1}^{t} \pi_i \max_i \sum_{j=1}^{t} q_{ij} \tag{4-20}$$

$$= \gamma \tag{4-21}$$

$$< 1. \tag{4-22}$$

Hence, $\forall n \geq 2$, $\pi^*_t \neq 0$, $\pi^*_{t_i} \geq 0 \; \forall i = 1, \ldots, t$, and $\varepsilon > 0$,

$$P(|\, x_n(\omega) - 1 \,| > \varepsilon) \quad = P(x_n(\omega) \neq 1) \tag{4-23}$$

$$= \pi^*_t Q^n \begin{bmatrix} 1 \\ \vdots \\ 1 \end{bmatrix} \tag{4-24}$$

Therefore,

$$\sum_{n=2}^{\infty} P(x_n(\omega) \neq 1) = \sum_{n=2}^{\infty} \pi_t^* Q^n \begin{bmatrix} 1 \\ \vdots \\ 1 \end{bmatrix} \tag{4-25}$$

$$= \sum_{n=2}^{\infty} \left\| \pi_t^* Q^n \begin{bmatrix} 1 \\ \vdots \\ 1 \end{bmatrix} \right\|_2 \tag{4-26}$$

$$\leq \sum_{n=2}^{\infty} \left\| \pi_t^* Q^n \right\|_2 \sqrt{t} \quad \text{(Cauchy-Schwarz Inequality)} \tag{4-27}$$

$$= \sum_{n=2}^{\infty} \left\| \pi_t^* Q^{n-1} Q \right\|_2 \sqrt{t} \tag{4-28}$$

$$= \sum_{n=2}^{\infty} \left\| \pi_t^* Q^{n-1} \right\|_1 \left\| \frac{\pi_t^* Q^{n-1}}{\| \pi_t^* Q^{n-1} \|_1} Q \right\|_2 \sqrt{t} \tag{4-29}$$

$$\leq \sum_{n=2}^{\infty} \left\| \pi_t^* Q^{n-1} \right\|_1 \gamma \sqrt{t}, \ 0 \leq \gamma < 1 \tag{4-30}$$

$$= \sum_{n=2}^{\infty} \left\| \pi_t^* Q^{n-2} \right\|_1 \left\| \frac{\pi_t^* Q^{n-2}}{\| \pi_t^* Q^{n-2} \|_1} Q \right\|_1 \gamma \sqrt{t} \tag{4-31}$$

$$\leq \sum_{n=2}^{\infty} \left\| \pi_t^* Q^{n-2} \right\|_1 \gamma^2 \sqrt{t} \tag{4-32}$$

$$\leq \sum_{n=2}^{\infty} \left\| \pi_t^* \right\|_1 \gamma^n \sqrt{t} \tag{4-33}$$

$$\leq \sum_{n=2}^{\infty} K \gamma^n \tag{4-34}$$

$$< \infty. \quad \blacksquare \tag{4-35}$$

This result also implies a stronger *complete convergence* (Hsu and Robbins, 1947; cf. *convergence everywhere*, Papoulis, 1984, p. 189).

As indicated by the preceding Markov analysis, convergence to a global optimum in the limit for evolutionary algorithms can be achieved with the simple conditions of a selection operator that always maintains the best solution in the population and a variation operator that can reach any state from any other state (or via a succession of intermediate states). Conversely, when relying solely on a variation operator that cannot iteratively reach all states (e.g., a typical crossover operator) or a selection procedure that has the potential to lose the best solution in the population at any generation (e.g., proportional selection), asymptotic global convergence may not be guaranteed. These are not the only conditions that can cause an evolutionary algorithm to fail to converge globally, but they are common in reported applications.

4.2.3 The Error of Minimizing Expected Losses in Schema Processing

The search for appropriate solutions may be viewed as a problem of allocating trials (i.e., searching) in a large state space. Rather than view this search in terms of evaluating complete solutions, the search can be viewed from the perspective of evaluating partial solutions (subsets of complete solutions) and attempting to bring together different subsets of disparate solutions that are evaluated with high quality. This requires a reasonable protocol for measuring the quality of a subset of a complete solution, a situation that is not always possible to achieve. For example, in a poker game, consider the question of assigning credit to individual cards in a straight. The value of the hand (the straight) only can be measured in terms of how well it compares to other hands as a whole. What is the value of a single card in a straight? The question is meaningless. Yet just this sort of question must be answered when trying to find "good" subsets of highly integrated solutions.

One approach to this problem is to assess the quality of a subset of a solution in terms of the average quality (or "fitness") that is attained across all possible substitutions in that portion of the solution which is not held constant. For instance, suppose that a candidate solution vector $\mathbf{x}$ to a particular problem was of length three, that each element in $\mathbf{x}$ could take on binary values $\{0, 1\}$, and that it was desired to determine the credit for the first component $x_1 = 1$. All vectors that match this case could be written as [1##] where the # symbol represents a wild card that matches either element in $\{0, 1\}$. The value of [1##] then is the average of the worth of [100], [101], [110], and [111]. This is essentially the strategy offered in Fisher (1930), where the fitness of a single gene in a complex genetic milieu was evaluated in terms of the average effects of other gene substitutions. Of course, averaging is a linear operator, and unless the overall fitness of a vector $\mathbf{x}$ is a linear function of its components, this method of assigning credit may be misleading.

Friedberg (1958) offered plans for assigning credit to subsets of entire solutions to given problems. Recall from Chapter 3 that Friedberg's efforts were aimed at using evolution to automatically generate a computer program and that programs which contained identical instructions were viewed as being in the same *class*. This notion was similar to the idea of "schemata" offered later by Holland (1973, 1975), which essentially offered the straightforward extension of Fisher's concept to "genes" comprising more than one element.

Consider the case where a solution is described using symbols from the set $\mathbf{A} = \{0, 1, \ldots, a\}$. A schema (singular of schemata) is a template of symbols from $\mathbf{A}$ and $\{\#\}$, where # is a "don't care" symbol that would match any symbol from $\mathbf{A}$. For example, if $\mathbf{A} = \{0, 1\}$, then the schema [1##0] includes [1000], [1010], [1100], and [1110]. There are many such sets of strings, and each may be viewed as a distribution from which samples may be drawn. The worth of any particular schema may be described as the average of the worth of all complete strings contained in the schema. Viewed from the obverse perspective, every complete solution offers partial information about the worth of sampling in every schema for which that solution is a member. Sampling from [1111] offers a datum concerning the payoffs that can be achieved from [1###], [#1##], [##1#], and so forth. This property was described as *intrinsic parallelism* in Holland (1975, p. 71) (also described as *implicit parallelism*), in that a single sample appears to intrinsically provide information regarding many schemata. Holland (1975, p. 71) speculated that bit strings would provide the optimal representation because the greatest number of schemata are created (and therefore the greatest intrinsic parallelism) when a problem is coded as a string of bits $\{0, 1\}$ (i.e., $|\mathbf{A}| = 2$). The sampling problem then is how best to allocate trials to alternative schemata in a population.

In order to optimally allocate trials a loss function describing the worth of each trial must be formulated. Holland (1975, pp. 75–83) considered and analyzed the problem of minimizing expected losses while sampling from alternative schemata. The essence of this problem can be modeled as a two-armed bandit (slot machine): Each arm of the bandit is associated with a random payoff described by a mean and variance (much like the result of sampling from a particular schema where the payoff depends on which instance of that schema is sampled). The goal is to best allocate trials to the alternative arms conditioned on the payoffs that have already been observed in previous trials. Holland (1975, pp. 85–87) extended the case of two competing schemata to any number of competing schemata. The results of this analysis were used to guide the formulation of one early version of evolutionary algorithms, so it is important to review this analysis and its consequences. It is particularly important because this formulation has been shown mathematically to be flawed (Rudolph, 1997a; Macready and Wolpert, 1998).

4.2.3.1   *The Two-Armed Bandit Problem.*   Holland (1975, pp. 75–83) examined the two-armed bandit problem in which there are two random variables, RV_1 and RV_2 (representing two slot machines) from which samples may be taken. The two random variables are assumed to be independent and possess some unknown means (μ_1, μ_2) and unknown variances (σ_1^2, σ_2^2). A trial is conducted by sampling from a chosen random variable; the result of the trial is the payoff. Suppose some number of trials has been devoted to each random variable (n_1 and n_2, respectively) and that the average payoff (the sum of the individual payoffs divided by the number of trials) from one random variable is greater than for the other. Let RV_{high} be the random variable for which the greater average payoff has been observed (not necessarily the random vari-

able with the greater mean), and let RV_{low} be the other random variable. The objective in this problem is to allocate trials to each random variable so as to minimize the expected loss function:

$$L(n_1, n_2) = [q(n_1, n_2)n_1 + (1 - q(n_1, n_2))n_2] \cdot |\mu_1 - \mu_2| \qquad (4\text{-}36)$$

where $L(n_1, n_2)$ describes the total expected loss from allocating n_1 samples to RV_1 and n_2 samples to RV_2, and

$$q(n_1, n_2) = \frac{\Pr(\bar{x}_1 > \bar{x}_2), \text{ if } \mu_1 < \mu_2}{\Pr(\bar{x}_1 < \bar{x}_2), \text{ if } \mu_1 > \mu_2} \qquad (4\text{-}37)$$

where $\bar{x}_1$ and $\bar{x}_2$ designate the mean payoffs (sample means) for having allocated n_1 and n_2 samples to RV_1 and RV_2, respectively. Note that specific measurements and the number of samples yield explicit values for the sample means.

As summarized in Goldberg (1989, p. 37), Holland (1975, pp. 77–78) proved the following (notation consistent with Holland, 1975):

THEOREM
Given N trials to be allocated to two random variables with means $\mu_1 > \mu_2$ and variances σ_1^2 and σ_2^2, respectively, and the expected loss function described in Eq. 4-36, the minimum expected loss results when the number of trials allocated to the random variable with the lower observed average payoff is

$$n^* \cong b^2 \ln\left[\frac{N^2}{8\pi b^4 \ln N^2}\right] \qquad (4\text{-}38)$$

where $b = \sigma_1/(\mu_1 - \mu_2)$. The number of trials to be allocated to the random variable with the higher observed average payoff is $N - n^*$.

If the assumptions associated with the proof (Holland, 1975, pp. 75–83) hold true, and if the mathematical framework were valid, the above analysis would apply equally well to a k-armed bandit as to a two-armed bandit.

Unfortunately, as offered in Macready and Wolpert (1998), the flaw in Holland (1975, pp. 77–85) stems from considering the unconditioned expected loss for allocating $N - 2n$ trials after allocating n to each bandit, rather than the expected loss conditioned on the information available after the $2n$ pulls. Macready and Wolpert (1998) showed that a simple greedy strategy which pulls the arm that maximizes the payoff for the next pull based on a Bayesian update from prior pulls outperforms the strategy offered in Holland (1975, p. 77). The tradeoff between exploitation (pulling the best Bayesian bandit) and exploration (pulling the other bandit in the hopes that the current Bayesian information is misleading) offered in Holland (1975) is not optimal for the problem studied. Macready and Wolpert (1998) remarked that the algorithms proposed in Holland (1975) "are based on the premise that one should engage in exploration, yet for the very problem invoked to justify genetic exploration, the strategy of not exploring at all performs better than $n*$ exploring algorithms."

4.2.3.2 Extending the Analysis for "Optimally" Allocating Trials.

Holland likened the problem of minimizing expected losses while sampling from k-armed bandits to the problem of minimizing expected losses while sampling from various schemata. From the previous analysis, it was proposed that to optimally allocate trials among the competing schemata in each solution, an exponential amount of attention (in terms of reproductive emphasis) should be devoted to the observed best solutions in order to minimize the expected losses while sampling over all schemata. Proportional selection was in turn chosen as a means of selecting parents based on observed fitness and on the belief that this method can allocate exponentially increasing trials to the observed best solutions (Goldberg, 1989, pp. 30–31).

It has been widely asserted (e.g., Holland, 1975, pp. 102–103, Davis, 1991, pp. 19–20) that if the number of copies of a solution comprising independent schemata is made proportional to its fitness (reproduction with emphasis), then the relative increase or decrease of the frequency of any schema, s, in that solution follows the relationship:

$$E(s) = (m_s f_s / f_m) - d \qquad (4\text{-}39)$$

where $E(s)$ is the expected number of solutions in the population containing s, m_s is the current number of solutions in the population containing s, f_s is the mean fitness of all solutions in the population that contain s, f_m is the average fitness of all solutions in the population, and d is a residual effect of "disruption" caused by crossover and mutation. This result is known as the *schema theorem*. "In effect, the schema theorem says that a schema occurring in a chromosome with above-average evaluations will tend to occur more frequently in the next generation, and one occurring with below-average evaluations will tend to occur less frequently (ignoring the effects of mutation and crossover)" (Davis, 1991, p. 20). In light of the k-armed bandit analysis in Holland (1975, pp. 77–85), the schema theorem was viewed as "fundamental" (Goldberg, 1989, p. 33), the key descriptor of why evolutionary algorithms can be successful: They optimally allocate trials to competing schemata in an implicitly parallel manner. In light of Macready and Wolpert (1998), there now appears to be no support for viewing the schema theorem as having fundamental importance. The theorem simply describes the expected number of each schemata at the next generation under proportional selection when each complete solution is assigned a specified fitness value.

4.2.3.3 Limitations of the Analysis.

Even when considering a correct analysis of the k-armed bandit problem and its relation to allocating samples

in an evolutionary algorithm, one important drawback to this "optimal allocation of trials" is that it only considers how to minimize expected losses when
independent samples are drawn from the random variables. Even if the
schemata were in fact random variables (in many cases they are not), most
evolutionary algorithms employ coding strings such that schemata are not
sampled independently (see Davis, 1985). Moreover, there may be very few
samples of any particular schema, lower than required for the central limit theorem arguments made to justify assumptions that payoffs follow normal distributions (as required in the derivation in Holland, 1975).

Moreover, the utility of minimizing expected losses in this sampling problem may be questioned. Such a criterion may preclude the discovery of a globally optimal solution. Consider two random variables with corresponding distributions that describe the worth of their various members of a given schema,
as depicted in Figure 4-3. The first distribution has mean μ_1 and variance σ_1^2,
whereas the second distribution has mean $\mu_2 < \mu_1$ and variance $\sigma_2^2 \gg \sigma_1^2$. In

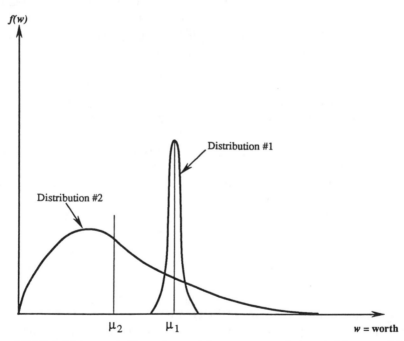

Figure 4-3 Minimizing expected losses does not always correspond to maximizing potential gains.
The illustration depicts two distributions of the fitness of various strings contained
within each schema (typically viewed as random variables). Although the mean of the
first distribution is greater than the second, strings with greater fitness reside in the second. To minimize expected losses, trials should be devoted to the first distribution,
which thereby precludes discovering the global best string.

order to minimize the expected loss (Eq. 4-36), all trials should be devoted to the first distribution because it has the larger mean. This in turn precludes the discovery of the best string, for it is a member of the second distribution. Identifying a schema with above-average performance does not, in general, provide information about which particular complete solution (which may be described by very many schemata) has the greatest fitness. The single solution with the greatest fitness (i.e., the globally optimal solution) may be described in schemata with below-average performance. The criterion of minimizing expected losses is quite conservative and may prevent successful optimization.

4.2.4 Misallocating Trials and the Schema Theorem in the Presence of Noise

The fitness associated with a particular schema H depends on which instances of that schema are evaluated. Moreover, in real-world practice, the evaluation of a particular string will often include some random effects (e.g., observation error). That is, the observed fitness of a particular schema H may not be described by a constant value, but rather by a random variable with an associated probability density function (pdf). Selection operates not on the mean of all possible samples of the schema H, but only on the observed fitness associated with each particular instance of the schema H in the population. It is therefore of interest to assess the effect that random observation error may impose on the allocation of trials to possible schemata and identify any bias in sampling that may occur as a result of introducing random effects. The principal question is whether or not reliable estimates of the average fitnesses of contending schemata are sufficient to propagate schemata according to the schema theorem (assuming that this particular allocation of trials is in fact desired).

Fogel and Ghozeil (1997a) have analyzed several cases of random fitness distributions under proportional selection. Neglecting the effects of variation operators, we find the expected number of each schema H in the population to be:

$$E\,P(H, t + 1) = P(H, t)\frac{f(H, t)}{\bar{f}_t} \tag{4-40}$$

where H is a particular schema (the notation of H is often used to denote the schema as a hyperplane), $P(H, t)$ is the proportion of strings in the population at time t that contain schema H, $f(H, t)$ is the mean fitness of strings in the population at time t that contain H, $\bar{f}_t$ is the mean fitness of all strings in the population at time t, and $E\,P(H, t + 1)$ denotes the expected proportion of strings in the population at time $t + 1$ that will contain the schema H. Suppose that $f(H, t)$ is a random variable. It then follows that $\bar{f}_t$ will also be a random variable. Fogel and Ghozeil (1997a) showed that the expected proportion of schema H at $t + 1$ is $E(Z)$ where:

$$Z = \frac{S}{S + S'}$$ (4-41)

and S and S' are the random variables describing the fitness of schema H and its complement H' (i.e., all schemata disjoint from H).

Unfortunately, $E(Z)$ cannot be calculated directly as the ratio of $E(S)$ to the sum of $E(S)$ and $E(S')$ because the expectation of a ratio of two random variables is not necessarily equal to the ratio of their expectations. It is of interest to identify the difference between the correct expression for the expectation of Z under proportional selection:

$$E\left(\frac{S}{S + S'}\right)$$ (4-42)

and the incorrect expression

$$\frac{E(S)}{E(S) + E(S')}$$ (4-43)

for this is the error that derives from ignoring the individual outcomes of each random variable and their associated probabilities, and instead treating all schemata only in terms of their mean fitness. Equation (4-42) can be determined using straightforward random variable transforms under given assumptions regarding the distributions of S and S'.

Fogel and Ghozeil (1997a) provided detailed analysis of cases where the fitnesses of schemata S and S' take on discrete or continuous probability mass or density functions. To provide a few examples here, consider the following:

EXAMPLE 1 For $S \sim U(0, 2\mu_S)$ and $S' \sim U(0, 2\mu_{S'})$:

$$E(Z) = \frac{1}{2}\left[\frac{\mu_{S'}}{\mu_S}\ln\left(\frac{\mu_{S'}}{\mu_S + \mu_{S'}}\right) - \frac{\mu_S}{\mu_{S'}}\ln\left(\frac{\mu_S}{\mu_S + \mu_{S'}}\right) + 1\right].$$ (4-44)

Figure 4-4 shows $E(Z)$ as a function of $\mu_{S'}$ (denoted by $E(S')$) for $\mu_S = 1$. The graph indicates that (4-43) equals (4-42) only when $\mu_S = \mu_{S'}$, as can be confirmed by inspection. This is a rather uninteresting condition, however, because when $\mu_S = \mu_{S'}$, the random variables for schema H and schema H' are identical. Thus, the decision of which schema to allocate more samples to is irrelevant. ∎

EXAMPLE 2 For $S \sim \exp(\lambda_S)$ and $S' \sim \exp(\lambda_{S'})$:

$$E(Z) = \frac{\lambda_S\lambda_{S'}}{(\lambda_S - \lambda_{S'})^2}\left[\ln\frac{\lambda_S}{\lambda_{S'}} - \frac{\lambda_S - \lambda_{S'}}{\lambda_S}\right],$$ (4-45)

for $\lambda_S \neq \lambda_{S'}$, and is 0.5 otherwise. ∎

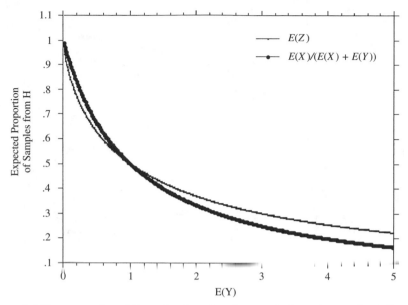

Figure 4-4 The expectation of Z as a function of $\mu_{S'}$ for $\mu_S = 1$ when schemata S and S' both follow uniform distributions. When $\mu_{S'}$ for μ_S (i.e., $\mu_{S'} = 1$ here), no difference between $\mu_S/(\mu_S + \mu_{S'})$ and $E(Z)$ is introduced; however, a bias is introduced for all other values of $\mu_{S'}$ (from Fogel and Ghozeil, 1997a).

Figure 4-5 shows $E(Z)$ as a function of $\lambda_{S'} = \mu_{S'}^{-1}$ for $\mu_S = 1$. Expression (4-43) is seen to equal (4-42) only when $\lambda_{S'} = \lambda_S$. Again, this is an uninteresting condition because the random variables for schema H and schema H' are identical when $\lambda_{S'} = \lambda_S$, and the decision of which schema to sample is of no consequence.

EXAMPLE 3   For $S \sim \gamma(1, c_S)$ and $S' \sim \gamma(1, c_{S'})$:

$$E(Z) = \frac{c_S}{c_S + c_{S'}}. \tag{4-46}$$

Therefore, these distributions on S and S' yield a case where (4-43) offers the same result as (4-42). ∎

EXAMPLE 4   For $S \sim N(0, 1)$ and $S' \sim N(0, 1)$, a case of Gaussian noise added as observation error, proportional selection is unworkable. The method can only be used for strictly positive values of S and S'. But even if this constraint were ignored, the density function of

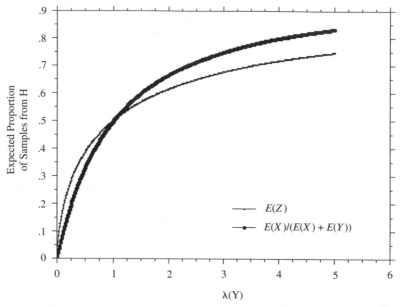

Figure 4-5 The expectation of Z as a function of $\lambda_{S'}$ for $\mu_{S'}^{-1} = 1$ when schemata S and S' both follow exponential distributions. When $\lambda_{S'} = \lambda_S$ (i.e., $\lambda_{S'} = 1$ here), no difference between $\mu_S/(\mu_S + \mu_{S'})$ and $E(Z)$ is introduced; however, a bias is introduced for all other values of $\lambda_{S'}$ (from Fogel and Ghozeil, 1997a).

$$f(z) = \left[\frac{\pi}{2}\left(1 + \left(\frac{z - 0.5}{0.5}\right)^2\right) \right]^{-1} \quad -\infty < z < \infty \qquad (4\text{-}47)$$

is a Cauchy random variable with scale parameters (½, ½). Since Z is Cauchy, the expectation of Z does not exist. ∎

The preceding results show that the use of proportional selection on observations from random variables, for all but one of the cases examined, leads to a biased sampling of schemata. The expected proportion of a particular schema H in the population at the next time step is not generally governed by the ratio of the mean of that schema H to the sum of the means of schema H and schema H'. Only for the case of specific gamma-distributed fitness scores did (4-43) match the result of (4-42). Reliably estimating mean hyperplane (schema) performance is not sufficient to predict the expected proportion of samples that will be allocated to a particular hyperplane under proportional selection. Thus, the schema theorem only applies to specific realized instances of schema fitness and cannot be used, in general, to assess the expected rates of schemata propagation over multiple iterations or when facing random initialization because the random effects that come into play may introduce a sampling bias (also see Fogel and Ghozeil, 1998 for other cases analyzed).

4.2.5 Analyzing Selection

There are many versions of selection in evolutionary algorithms. Each generally attempts to put more emphasis on those solutions in the population that are of higher quality, and correspondingly less emphasis on those that are of lower quality. This emphasis can be absolute (e.g., only maintain some percentage of the best-scoring individuals and eliminate the rest) or qualified (e.g., provide a greater probability of reproducing a solution with higher relative quality). The stringency of selection can be measured in terms of (1) distribution of fitness (error) scores in the population before and after selection, and (2) statistics that are related to this distribution. Most commonly, attention has been given to the expected number of generations that are required for the best solution to replicate and completely dominate all instances in the population, termed the *takeover time* (Goldberg and Deb, 1991; Bäck, 1994; Fogel and Fogel, 1995).

Bäck (1994) provided a comprehensive comparison of the takeover times of several common versions of selection including: (1) proportional selection, (2) two forms of tournament selection, (3) linear ranking selection, and (4) truncation selection. To review the methods, proportional selection acts on positive fitness values only and replicates each individual in the population with a probability equal to the ratio of its fitness to the sum of the fitnesses of all members of the population. The two forms of tournament selection either (1) iteratively select subsets of size q from the population and choose the best individual among the q to be a parent, or (2) compete each individual against q competitors and assign points based on pairwise comparisons, with selection then applied based on each individual's number of points rather than their raw fitness. Linear ranking selection sorts all individuals according to their fitness and then assigns a probability of selecting the ith individual as

$$\Pr(i\text{th individual selected}) = \frac{\left(\eta^+ -(\eta^+ - \eta^-)\frac{i - 1}{\lambda - 1}\right)}{\lambda} \tag{4-48}$$

where λ is the number of offspring, and $1 \leq \eta^+ \leq 2$ and $\eta^- = 2 - \eta^+$. Finally, in the case of μ parents creating λ offspring, truncation selection chooses only the best μ individuals from the union of parents and offspring, or simply from the offspring.

Selection pressure may be defined as being stronger with decreasing takeover time. Under this definition, Bäck (1994) showed, for typical parameter settings, that the degree of selection pressure increases in the order of proportional selection, linear ranking, tournament selection, and truncation selection. Tournament selection can be tuned by varying the parameter q to be small or large. As $q \to \infty$, tournament selection approaches truncation selection. The greater the selection pressure, the faster the population will converge. If there are multiple minima or maxima to overcome, however, this faster convergence may not be desirable because it may be toward only a local optimum. Under certain conditions, it will also be more difficult to escape from a local optimum when selection pressure is stronger rather than weaker (see Beyer and Fogel, 1997 for mathematical analysis).

4.2.6 Convergence Rates for Evolutionary Algorithms

Although asymptotic global convergence of evolutionary algorithms is a significant result, it is far more important to assess the rate of convergence in specific problems. The most insightful theoretical analyses of convergence rates in evolutionary algorithms have been contributed by Rechenberg (1973), Schwefel (1981), Bäck et al. (1991), Bäck et al. (1993), Beyer (1994, 1995), and Rudolph (1994b, 1996). Many of these results have been aimed at algorithms that operate on continuous parameters and employ Gaussian variations to single parents, and potentially discrete and intermediate (averaging) recombination as well.

One of the simplest evolutionary algorithms for real-valued continuous optimization problems, described as a $(1 + 1)-\text{EA}$, starts with a single parent, $\mathbf{x}$, creates a single offspring, $\mathbf{x}'$, from this parent by imposing a multivariate Gaussian perturbation with mean zero and standard deviation σ to the parent, and then selects the better of the two trial solutions as the parent for the next iteration.

The same standard deviation σ is applied to each component of the vector $\mathbf{x}$ during mutation. For some problems, the variation of σ (i.e., the step-size control parameter in each dimension) can be computed to yield an optimal rate of convergence. Let the convergence rate be defined as the ratio of the Euclidean distance covered toward the optimum solution to the number of trials required to achieve the improvement. Rechenberg (1973) calculated the convergence rates for two functions:

$$f_1(\mathbf{x}) = F(x_1) = c_0 + c_1 x_1, \mathsf{A}\, i \in \{2, \ldots, n\} : -b/2 \le x_i \le b/2 \qquad (4\text{-}49)$$

$$f_2(\mathbf{x}) = \Sigma\, x_i^2 \qquad (4\text{-}50)$$

where $\mathbf{x} = (x_1, \ldots, x_n)^{\mathrm{T}} \in \mathfrak{R}^n$, where to remain consistent with notation in Rechenberg (1973), n is now the number of dimensions. Function f_1 is termed the *corridor model* and represents a linear function with inequality constraints. Improvement is accomplished by moving along the first axis of the search space inside a corridor of width b. Function f_2 is termed the *sphere model* and is a simple n-dimensional quadratic bowl.

The expected rates of convergence, φ_1 and φ_2, for f_1 and f_2 (Rechenberg, 1973) are

$$\varphi_1 = \sigma(2\pi)^{-0.5}\left[1 - \left(\frac{2}{\pi}\right)^{0.5}\left(\frac{\sigma}{b}\right)\right]^{n-1}, n \gg 1 \tag{4-51}$$

$$\varphi_2 = \sigma(2\pi)^{-0.5}\exp\left[-\left(\frac{n\sigma}{r\sqrt{8}}\right)\right]^2 - \sigma(2\pi)^{-0.5}\left(\sqrt{\pi}\left[\frac{n\sigma}{r\sqrt{8}}\right]\left\{1 - \text{erf}\left[\frac{n\sigma}{r\sqrt{8}}\right]\right\}\right), n \gg 1 \tag{4-52}$$

where erf(x) refers to the error function and r refers to the distance from the optimum solution. By taking the derivatives of φ_1 and φ_2 with respect to σ, setting them equal to zero, and solving for σ, the optimum settings for the standard deviation to use, and the optimum rates of convergence, can be calculated as

$$\sigma_1^{\text{opt}} = \left(\frac{\pi}{2}\right)^{0.5}\left(\frac{b}{n}\right); \quad \varphi_1^{\max} = (2e)^{-1}\left(\frac{b}{n}\right) \tag{4-53}$$

$$\sigma_2^{\text{opt}} = 1.224\left(\frac{r}{n}\right); \quad \varphi_2^{\max} = 0.2025\left(\frac{r}{n}\right) \tag{4-54}$$

The optimum convergence rate is obtained for both functions when the standard deviation for the multivariate Gaussian perturbation is inversely proportional to the dimension n. Furthermore, the maximum rate of convergence is also inversely proportional to n.

Given the optimum standard deviations, the optimum probability of generating a successful mutation can be calculated as

$$p_1^{\text{opt}} = (2e)^{-1} \approx 0.184 \tag{4-55}$$

$$p_2^{\text{opt}} \approx 0.270. \tag{4-56}$$

Rechenberg (1973), noting the similarity of these two values, suggested the following rule:

> *The ratio of successful mutations to all mutations should be $\frac{1}{5}$. If this ratio is greater than $\frac{1}{5}$, increase the variance; if it is less, decrease the variance.*

Schwefel (1981) suggested measuring the success probability on-line over $10n$ trials (where there are n dimensions) and adjusting σ at iteration t by

$$\sigma(t) = \begin{cases} \sigma(t-n)\cdot\delta, & \text{if } p_s < 0.2 \\ \sigma(t-n)/\delta, & \text{if } p_s > 0.2 \\ \sigma(t-n), & \text{if } p_s = 0.2, \end{cases} \tag{4-57}$$

with $\delta = 0.85$ and p_s equaling the number of successes in $10n$ trials divided by $10n$, which yields convergence rates of geometric order for both f_1 and f_2 (Bäck et al., 1993; see Bäck, 1996 for corrections to the update rule offered in Bäck et al., 1993).

Schwefel (1977, 1981) introduced two truncation-based evolutionary algorithms, described by $(\mu + \lambda)-$EA and $(\mu, \lambda)-$EA. In the former, μ parents create λ offspring, and the best μ solutions from among all $\mu + \lambda$ are selected to become parents of the next generation. In the latter, the best μ solutions are selected only from the λ offspring. Each solution is characterized not only by an n-dimensional vector of object variables $\mathbf{x}$, but also by as many as $n(n + 1)/2$ additional mutation parameters, which may include up to n different variances $c_{ii} = \sigma_i^2$ $(i \in \{1, \ldots, n\})$ as well as up to $n(n - 1)/2$ covariances c_{ij} $(i \in \{1, \ldots, n - 1\}, j \in \{i + 1, \ldots, n\}, c_{ij} = c_{ji})$ describing an n-dimensional Gaussian random vector with probability density function:

$$p(\mathbf{z}) = \frac{\exp(-0.5\mathbf{z}'C^{-1}\mathbf{z})}{\sqrt{(2\pi)^n |C|}}.$$
(4-58)

For convenience in ensuring that C^{-1} remains positive definite, the algorithm operates on equivalent rotation angles α_l $(0 \le \alpha_l < 2\pi)$, rather than directly on the coefficients c_{ij}. The algorithm is cumbersome to explicate, but in pseudocode it becomes:

> $t = 0$;
> *initialize* $P(0) = \{a_1(0), \ldots, a_\mu(0)\}$
>   where $a_i(0) = (\mathbf{x}_i(0), \sigma_i(0), \alpha_i(0))$,
> $\mathbf{x}_i(0) \in \Re^n, \sigma_i(0) \in \Re^{+n}, \alpha_i(0) \in [0,2\pi)^{n(n-1)/2}, i = 1, \ldots, \mu$;
> *evaluate* $P(0)$;
> **while** *termination criteria not fulfilled* **do**
>   *recombine:* $a_i'(t) = \mathbf{r}(P(t)), i = 1, \ldots, \lambda$;
>   *mutate:* $a_i''(t) = \mathbf{m}(a_i'(t)), i = 1, \ldots, \lambda$;
>   *evaluate:* $P'(t) = \{a_1''(t), \ldots, a_\lambda''(t)\}$;
>   /* yields $\{f(\mathbf{x}_1''(t)), \ldots, f(\mathbf{x}_\lambda''(t))\}$ */
>   *select* $P(t + 1) =$
> $\mathbf{s}_{(\mu,\lambda)}(P'(t))$ if $(\mu, \lambda)-$ES
> $\mathbf{s}_{(\mu+\lambda)}(P'(t) \cup P(t))$ if $(\mu + \lambda)-$ES;
> $t = t + 1$;
> **od**

The mutation operator is extended (without referring to the time counter t) as

$$m(a_k) = a_k' = (\mathbf{x}', \sigma', \alpha')$$
(4-59)

with each operation as follows:

$$\sigma_i' = \sigma_i \cdot \exp(\tau_0 \cdot \Delta\sigma_0) \cdot \exp(\tau \cdot \Delta\sigma_i)$$
(4-60)

$$\alpha_j' = \alpha_j + \beta \cdot \Delta\alpha_j$$
(4-61)

$$x_i' = x_i + z_i(\sigma', \alpha').$$
(4-62)

with $i = 1, \ldots, n, j = 1, \ldots, n(n - 1)/2$.

The mutations of the object variables, $\mathbf{x}$, are correlated according to the pairwise rotation vector α' and scaling σ'. Alterations $\Delta\sigma_0$, $\Delta\sigma_i$, and $\Delta\alpha_j$ are again standard Gaussian random variables, and the constants $\tau \propto \left(\sqrt{2\sqrt{n}}\right)^{-1}$, $\tau_0 \propto \left(\sqrt{2n}\right)^{-1}$ and $\beta \approx 0.0873$ (5°) as suggested by Bäck et al. (1993) are robust to diverse response surfaces. $\Delta\sigma_0$ is a global scaling that affects all σ_i, $i \in \{1, \ldots, n\}$, while $\Delta\sigma_i$ is recalculated for each $i \in \{1, \ldots, n\}$ allowing for individual changes to the mean step sizes σ_i.

Recombination can be implemented in many ways. Bäck et al. (1993) suggest the following methods, which are described for object variables but can also be applied to mutation strategy variables:

$$x'_i = \begin{cases} x_{S,i} & (1) \\ x_{S,i} \text{ or } x_{T,i} & (2) \\ x_{S,i} + u(x_{T,i} - x_{S,i}) & (3) \\ x_{S_i,i} \text{ or } x_{T_i,i} & (4) \\ x_{S_i,i} + u_i(x_{T_i,i} - x_{S_i,i}) & (5), \end{cases} \qquad (4\text{-}63)$$

where S and T denote two arbitrary parents, u is a uniform random variable over the interval $[0, 1]$, and the different variations are: (1) no recombination, (2) discrete recombination, (3) intermediate recombination, and the global versions (4) and (5) of (2) and (3), respectively, in which any two parents can contribute to each x'_i. The utility of these operations is highly problem dependent.

The preceding algorithms can be shown to have geometric convergence rates on certain response surfaces (Bäck et al., 1993). The actual rate varies by problem, and appropriate analysis requires restricting attention to specific surfaces (e.g., strongly convex surfaces). Consider the following definitions with notation from Bäck et al. (1993):

DEFINITION 1   A random vector $\mathbf{z}$ of dimension n is said to possess an *elliptical distribution* iff it has stochastic representation $\mathbf{z} = rA'\mathbf{u}$, where the random vector $\mathbf{u}$ is uniformly distributed on a hypersphere surface of dimension n, stochastically independent to a nonnegative random variable r and where the matrix $A : k \times n$ with rank$(A'A) = k$. If $A : n \times n$ and $A = I$, then $\mathbf{z}$ is said to possess a *spherical (symmetric) distribution*. If $\mathbf{z}$ is used to describe the mutation probability density function and is chosen to be a multivariate normal random variable with zero mean and covariance $C = \sigma^2 I$, then the step size r has a central $\chi_n(\sigma)$-distribution (Fang et al., 1990; Bäck et al., 1993). If the matrix A and the distribution of r are fixed during the optimization, the evolution strategy may be said to have a *stationary* step-size distribution; otherwise the distribution is described as being *adaptive*.

DEFINITION 2   The value $\delta_t \equiv E[f(X_t) - f^*]$, where X_t is the stochastic process defining the sequence of trial solutions and f^* is the minimum value of the function f, is said to be the *expected error* at step t. An algorithm has a *sublinear convergence* rate iff $\delta_t = O(t^{-b})$ with $b > 0$ and a *geometric convergence rate* iff $\delta_t = O(r^t)$ with $r \in (0, 1)$.

DEFINITION 3   A function $f : \Re^n \to \Re$ is said to be (l, Q)-*strongly convex* iff it is continuously differentiable and with some constants $l > 0, Q \geq 1$ there holds

$$l \cdot \|\mathbf{x} - \mathbf{y}\|^2 \leq (\nabla f(\mathbf{x}) - \nabla f(\mathbf{y}))'(\mathbf{x} - \mathbf{y}) \leq Ql \cdot \|\mathbf{x} - \mathbf{y}\|^2$$

for all $\mathbf{x}, \mathbf{y} \in M_F$, where ∇f denotes the gradient of f and M_F denotes the feasible range for $\mathbf{x}$ and $\mathbf{y}$.

The following theorem can now be stated by applying the above definitions:

THEOREM
Let f be a (l, Q)-strongly convex function and $\mathbf{z} \equiv r\mathbf{u}$, where r has nonvoid support $(0, a) \subseteq \Re$. Then the expected error of a $(1 + 1)-\text{EA}$ decreases for any starting point $\mathbf{x}_0 \in M_F$ with the following rates:

$$E[f(X_t) - f^*] \leq \begin{cases} O(t^{-2/n}), & \text{stationary step-size distribution} \\ O(\beta^t), & \text{adapted step-size distribution} \end{cases}$$

with $\beta \in (0, 1)$.

To ensure geometric convergence, the step sizes R can be adjusted by choosing $R(t + 1) = \zeta \|\nabla f(\mathbf{x}(t))\| \cdot R(t)$, $\zeta > 0$. Alternatively, the procedure offered by Rappl (1984), in which the step size is decreased by a factor $\gamma_1 \in (0, 1)$ if the trial solution is worse than the parent and increased by a factor $\gamma_2 > 1$ if the trial solution is better than the parent, can be guaranteed to yield geometrical convergence for $\gamma_1 \cdot \gamma_2 > 1$ (Bäck et al., 1993).

Bäck et al. (1993) provided a simple example. Let $f(\mathbf{x}) = \|\mathbf{x}\|^2$ with $M_F = \Re^n$. It is clear then that f is $(2, 1)$-strongly convex:

$$(\nabla f(\mathbf{x}) - \nabla f(\mathbf{y}))'(\mathbf{x} - \mathbf{y}) = 2\|\mathbf{x} - \mathbf{y}\|^2. \tag{4-64}$$

If $\mathbf{z}$, the mutation vector, is Gaussian with zero mean and covariance $C = \sigma^2 I$, the distribution of objective values follows:

$$f(\mathbf{x}(t) + \mathbf{z}(t)) \sim \sigma^2 \chi_n^2(\kappa), \tag{4-65}$$

where $\chi_n^2(\kappa)$ is a noncentral χ^2-distribution with n degrees of freedom and noncentrality parameter $\kappa = \|\mathbf{x}(t)\|^2/\sigma^2$. By Johnson and Kotz (1970, p. 135):

$$\frac{\chi_n^2(\kappa) - (n + \kappa)}{\left(2(n + 2\kappa)\right)^{0.5}} \to N \sim N(0, 1) \tag{4-66}$$

for $n \to \infty$. The relative variation of the objective function, V, can therefore be approximated as

$$V \equiv \frac{f\left(\mathbf{x}(t)\right) - f\left(\mathbf{x}(t + 1)\right)}{f\left(\mathbf{x}(t)\right)} \tag{4-67}$$

$$\sim 1 - \frac{\sigma^2}{\|\mathbf{x}(t)\|^2} \chi_n^2(\kappa) \tag{4-68}$$

$$\sim 1 - \frac{\sigma^2}{\|\mathbf{x}(t)\|^2}\left(n + \kappa + (2n + 4\kappa)^{0.5}N\right) \tag{4-69}$$

$$= -\frac{s^2}{n} - \frac{s^2}{n}\left(\frac{2}{n} + \frac{4}{s^2}\right)^{0.5}N \tag{4-70}$$

$$\approx -\frac{s^2}{n} - \frac{2s}{n}N \tag{4-71}$$

where $\sigma = s\,\|\mathbf{x}(t)\|/n$. Because the $(1 + 1)-$EA only accepts improvements, attention can be restricted to the expectation of the random variable $V^+ = \max\{0, V\}$:

$$E[V^+] = \int_0^\infty \frac{nu}{2s\sqrt{2\pi}} \exp\left[-0.5\left(\frac{nu + s^2}{2s}\right)^2\right] du \tag{4-72}$$

$$= \frac{1}{n}\left(s\left(\frac{2}{\pi}\right)^{0.5}\exp\left(\frac{-s^2}{8}\right) - s^2\left[1 - \Phi\left(\frac{s}{2}\right)\right]\right) \tag{4-73}$$

where u is a variable of integration and $\Phi(.)$ denotes the cumulative distribution function (c.d.f.) of a standard normal random variable. The expectation is maximized for $s* = 1.224$ (see Figure 4-6) such that $E[V^+] = 0.405/n$ and $\sigma* = (1.224/n)\,\sqrt{f(\mathbf{x})} = (0.612/n)\|\nabla f(\mathbf{x})\|$. The latter description is more general and will hold for any $f(\mathbf{x}) = \|\mathbf{x} + \mathbf{c}_1\|^2 + c_2$, where $\mathbf{c}_1 \in \mathfrak{R}^n, c_2 \in \mathfrak{R}$.

This convergence rate can be improved by generating multiple solutions from each parent. Schwefel (1977, 1981) considered the convergence properties of the $(1 + \lambda)-$EA. First, λ independent trials are generated from the best solution of the previous generation, and then the best of those trials is taken

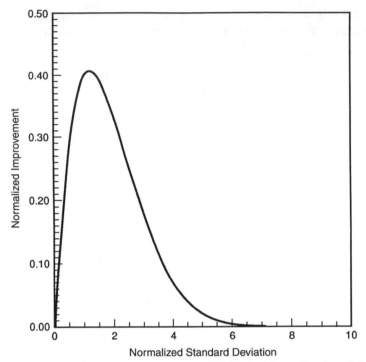

Figure 4-6 The normalized expected improvement $nE[V^+]$ plotted as a function of the normalized standard deviation $s = n\sigma/f(\mathbf{x}_t)^{0.5}$. The expected improvement is maximized for $s^* = 1.224$ (after Bäck et al., 1993).

to be the new parent if it outperforms the previous parent. The relative variation of the objective function value for a single trial is

$$V \approx \frac{-s^2}{n} - \frac{2s}{n}N \equiv \frac{-s^2}{n} + \frac{2s}{n}N \tag{4-74}$$

from above. The c.d.f. of V is $\Phi((x - \theta)/\eta)$, with $\theta = -(s^2/n)$ and $\eta = (2s/n)$. The distribution of the maximal relative variation of λ trials, V_λ, is given by

$$F_{V_\lambda} \equiv \Phi^\lambda\left[\frac{x - \theta}{\eta}\right], \tag{4-75}$$

because the λ trials are i.i.d. The expectation of V_λ is

$$E[V_\lambda] = \frac{1}{\eta}\int_{-\infty}^{\infty} ug\left(\frac{u - \theta}{\eta}, \lambda\right)du \tag{4-76}$$

where

$$g(x,\lambda) \equiv \frac{d\Phi^\lambda(x)}{dx} = \frac{\lambda}{\sqrt{2\pi}} \exp\left(\frac{-x^2}{2}\right)\Phi^{\lambda-1}(x) \qquad (4\text{-}77)$$

Let the random variable $V_\lambda^+ \equiv \max\{0,V_\lambda\}$ describe the relative improvement after λ trials. The expectation

$$E[V_\lambda^+] = \frac{1}{\eta}\int_0^\infty ug\left(\frac{u-\theta}{\eta}, \lambda\right)du \qquad (4\text{-}78)$$

Substituting $v = (u-\theta)/\eta$ yields

$$E[V_\lambda^+] = \int_{\frac{-\theta}{\eta}}^\infty (v\eta + \theta)\, g(v, \lambda)dv \qquad (4\text{-}79)$$

$$= \frac{2s}{\eta}\int_{J\nu z}^\infty v\, g(v, \lambda)dv - \frac{s^2}{\mu}\left[1 - \Phi^\lambda\left(\frac{s}{2}\right)\right]. \qquad (4\text{-}80)$$

Unfortunately, no closed-form expression for this integral appears to be available. But the optimal normalized standard deviations and improvements can be computed numerically (Figure 4-7). As seen in Figure 4-7b, the normalized improvement increases at what appears to be a logarithmic rate (Bäck et al., 1993).

Schwefel (1977, 1981) and Scheel (1985) have examined the rates of convergence for $(1, \lambda)-$EA on the same class of functions and have also identified the expected increase in convergence based on λ. The expected relative improvement of λ trials is the maximal relative variation V_λ. Substituting $v = (u-\theta)/\eta$ into Eq. 4-76 yields

$$E[V_\lambda] = \int_{-\infty}^\infty (v\eta + \theta)\, g(v, \lambda)dv \qquad (4\text{-}81)$$

$$= \frac{2s}{n}\int_{-\infty}^\infty v\, g(v, \lambda)dv - \frac{s^2}{n} \qquad (4\text{-}82)$$

$$= \frac{2s\, c_{1,\lambda} - s^2}{n} \qquad (4\text{-}83)$$

where

$$c_{1,\lambda} \equiv \int_{-\infty}^\infty v\, g(v, \lambda)dv \qquad (4\text{-}84)$$

denotes the expectation of the maximum of λ i.i.d. standard Gaussian random variables.

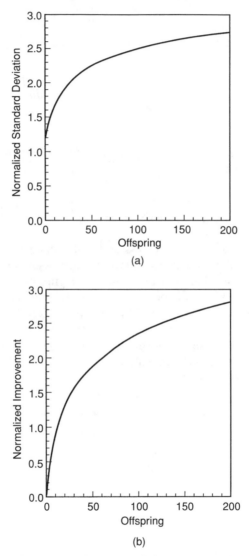

Figure 4-7 (a) The optimal normalized standard deviation plotted as a function of the number of offspring. (b) The optimal normalized improvement $n E[V^+]$ plotted as a function of the number of offspring. As only the outcome of the best trial is of relevance here, when more trials are conducted within each generation they may be performed with more risk, corresponding to a larger step size. Both plots are from Bäck et al. (1993).

Differentiating Eq. 4-83 with respect to s yields the optimum normalized standard deviation $s^* = c_{1,\lambda}$ so that the optimal standard deviation is

$$\sigma^* = (c_{1,\lambda}/2n)\|\nabla f(x)\| \tag{4-85}$$

for large n. The expected improvement is then $E[V] = \dfrac{c_{1,\lambda}^2}{n}$ which can be described asymptotically.

Let $X_i \sim N(0, 1)$ for $i = 1, \ldots, \lambda$. Then $M_\lambda \equiv \max\{X_1, \ldots, X_\lambda\}$ is a random variable with c.d.f. $P(M_\lambda \le x) = \Phi^\lambda(x)$. Consequently, $c_{1,\lambda} = E[M_\lambda]$. The normalized random variable $(M_\lambda - b_\lambda)/a_\lambda$ converges weakly to the random variable Y, which has Gumbel distribution $G(y) = \exp(-e^{-y})$. The normalizing constants are

$$a_\lambda = (2 \log \lambda)^{-0.5} \tag{4-86}$$

$$b_\lambda = (2 \log \lambda)^{-0.5} - \frac{\log \log \lambda + \log (4\pi)}{2\sqrt{2 \log \lambda}}. \tag{4-87}$$

Thus,

$$E[(M_\lambda - b_\lambda)/a_\lambda] \to E[Y] = \gamma \approx 0.577. \ldots, \tag{4-88}$$

where γ denotes Euler's constant. Continuing,

$$E[M_\lambda] \to b_\lambda + \gamma \cdot a_\lambda \tag{4-89}$$

$$= \sqrt{2 \log \lambda} - \frac{2\gamma - \log \log \lambda - \log (4\pi)}{2\sqrt{2 \log \lambda}} \tag{4-90}$$

$$\approx \sqrt{2 \log \lambda}, \tag{4-91}$$

and the expected optimal relative improvement becomes

$$E^*[V_\lambda] \approx (2 \log \lambda)/n \tag{4-92}$$

for large n and large λ. This asymptotic expression also holds for the $(1 + \lambda)-$EA and lends support to the apparent logarithmic increase exhibited in Figure 4-7b.

This result enables the computation of the relative speedup of processing λ trials in parallel [the $(1, \lambda)-$EA] over processing a single new trial [the $(1 + 1)-$EA]. Let

$$\beta \equiv E\left[\frac{f(\mathbf{x}(t)) - f(\mathbf{x}(t + 1))}{f(\mathbf{x}(t))} \right] \tag{4-93}$$

$$= \begin{cases} 0.405/n, & \text{if } (1 + 1)-\text{EA} \\ (2 \log \lambda)/n, & \text{if } (1, \lambda)-\text{EA}. \end{cases} \tag{4-94}$$

The expected error at step t is then:

$$E[f(x(t)) - f^*] = (1 - \beta)(f(\mathbf{x}(t-1)) - f^*) \tag{4-95}$$

$$= (1 - \beta)^t(f(\mathbf{x}(0)) - f^*) \tag{4-96}$$

$$= (1 - \beta)^t \delta_0. \tag{4-97}$$

Let $E[t_1]$ and $E[t_\lambda]$ denote the expected number of steps required to achieve an error of ε for a $(1 + 1)-$EA and a $(1, \lambda)-$EA, respectively. Then the relative increase in speed, S_λ, satisfies:

$$S_\lambda \equiv E[t_1]/E[t_\lambda] \tag{4-98}$$

$$= \frac{\log (\varepsilon/\delta_0)}{\log (1 - 0.405/n)} \frac{\log\left(1 - \dfrac{c_{1,\lambda}^2}{n}\right)}{\log (\varepsilon/\delta_0)} \tag{4-99}$$

$$= \frac{\log (1 - 2\log(\lambda)/n)}{\log (1 - 0.405/n)} \tag{4-100}$$

$$\approx \frac{2\log(\lambda)/n}{0.405/n} \tag{4-101}$$

$$= O(\log \lambda). \tag{4-102}$$

Thus, the relative increase in speed for using λ offspring is logarithmic.

4.2.7 Does a Best Evolutionary Algorithm Exist?

It is natural to ask if there is a best evolutionary algorithm that would always give superior results across the possible range of problems. Is there some choice of variation operators and selection mechanisms that will always outperform all other choices regardless of the problem? The answer is no: There is no best evolutionary algorithm. In mathematical terms, let an algorithm a be represented as a mapping from previously unvisited sets of points to a single new (i.e., previously unvisited) point in the state space of all possible points (solutions). Let $P(d_m^y|f, m, a)$ be the conditional probability of obtaining a particular sample d_m that yields a value y when algorithm a is iterated m times on cost function f. Given these preliminaries, Wolpert and Macready (1997) proved the so-called no free lunch theorem:

THEOREM
(No Free Lunch): For any pair of algorithms a_1 and a_2

$$\sum_f P(d_m^y \mid f, m, a_1) = \sum_f P(d_m^y \mid f, m, a_2)$$

(see Appendix A of Wolpert and Macready, 1997 for the proof).[3] That is, the sum of the conditional probabilities of obtaining each value d_m^y is the

---

[3] English (1996) showed that a similar no free lunch result holds whenever the values assigned to points are independent and identically distributed random variables.

same over all cost functions f regardless of the algorithm chosen. The immediate corollary of this theorem is that for any performance measure $\Phi(d_m^y)$, the average over all f of $P(\Phi(d_m^y)|f, m, a)$ is independent of a. In other words, there is no best algorithm, whether or not that algorithm is "evolutionary." Moreover, whatever an algorithm gains in performance on one class of problems is necessarily offset by that algorithm's performance on the remaining problems.

This simple theorem has been the source of much controversy in the field of evolutionary computation and some misunderstanding. In fact, some have remarked that this result has set the field of evolutionary computation back several years, although such comments cannot be taken seriously since no scientific fact should ever be viewed as a step backward. The frustration reflected in the comment, however, is genuine. A great deal of effort has been expended in finding the "best" set of parameters and operators for evolutionary algorithms since at least the mid-1970s. These efforts have involved the type of its combination, the probabilities for crossover and mutation, the representation, the population size, and so forth. Most of this research has involved empirical trials on benchmark functions, as will be described in the following section. But the no free lunch theorem essentially dictates that the conclusions made on the basis of such sampling are in the strict mathematical sense limited to those functions studied. Efforts to find the best crossover rate, the best mutation operator, and so forth, in the absence of restricting attention to a particular class of problems are futile.

For an algorithm to perform better than even random search (which is simply another algorithm), it must reflect something about the structure of the problem it faces. By consequence, it mismatches the structure of some other problem. Note, too, that it is not enough simply to state that a problem has some specific structure associated with it: that structure must be appropriate to the algorithm at hand. Moreover, the structure must be specific. It is not enough to say, as is often heard, "I am concerned only with real-world problems, not all possible problems, and therefore the no free lunch theorem does not apply." What is the structure of "real-world" problems? Indeed, what is a real-world problem? The obvious vague quality of this description is immediately problematic. What constitutes a real-world problem now might not have been a problem at all, say, 100 years ago (e.g., what to watch on television on a Thursday night). Nevertheless, simply narrowing the domain of concern without identifying the correspondence between the set of problems considered and the algorithm at hand does not suffice to claim any advantage for a particular method of problem solving.

Furthermore, the correspondence must be mathematical, not verbal (which is synonymous here with "vague"). For example, it is not sufficient to claim that "evolutionary algorithms are based on natural evolution, and I am concerned with problems from nature; therefore evolutionary algorithms are superior to other methods." Many algorithms are based on natural phenomena: (1) steepest descent methods are based on energy minimization as found in physical systems, (2) simulated annealing is based on the natural microscopic rearrangement processes modeled by statistical mechanics, and so forth. But it is well known that these procedures have vastly different optimization capabilities. In fact, there is evidence that runs counter to the above claim. Nature is often characterized by chaos and fractals, yet the limited available results using evolutionary algorithms on fractal problems are not especially encouraging (e.g., Bäck, 1996, p. 158), and chaotic systems by definition will be potentially resistant to any robust solution because of their sensitivity to initial conditions. Care must be taken to eschew simple reasoning that attempts to circumvent or dismiss the no free lunch theorem.

One apt example of how the match between an algorithm and the problem can be exploited was offered in De Jong et al. (1995). For a very simple problem of finding the two-bit vector **x** that maximizes the function:

$$f(\mathbf{x}) = \text{integer}(\mathbf{x}) + 1$$

where integer(**x**) returns 0 for [00], 1 for [01], 2 for [10], and 3 for [11], De Jong et al. (1995) employed an evolutionary algorithm with (1) one-point crossover at either a probability of 1.0 or 0.0, (2) a constant mutation rate of 0.1, and (3) a population of size five. In this trivial example, it was possible to calculate the exact probability that the global best vector would be contained in the population as a function of the number of generations. Figure 4-8 shows that, in this case, the use of crossover definitely increases the likelihood of discovering the best solution and that mutation alone is better than random search.

In this example the function f assigns values $\{1, 2, 3, 4\}$ to the vectors $\{[00],$ [01], [10], [11]\}, respectively, but this is not the only way to assign these fitness values to the possible strings. In fact, 4! = 24 different permutations could be used. De Jong et al. (1995) showed that the performance obtained with the 24 different permutations fall into three equivalence classes, each containing eight permutations that produce identical probability of success curves. Figure 4-8 shows the first equivalence class above, but Figure 4-9 shows the second and third equivalence classes, where the use of crossover is seen to be detrimental to the likelihood of success, and random search can outperform evolutionary search for the first 10 to 20 generations in the third equivalence class. In the first equivalence class, crossover can combine the second- and third-best vectors to generate the best vector. In the second and third equivalence classes, it cannot usefully combine these vectors: The structure of the problem does not match the structure of the crossover operator. Any specific search operator can be rendered superior or inferior simply by changing the structure of the problem.

Intrinsic to every evolutionary algorithm is a representation for manipu-

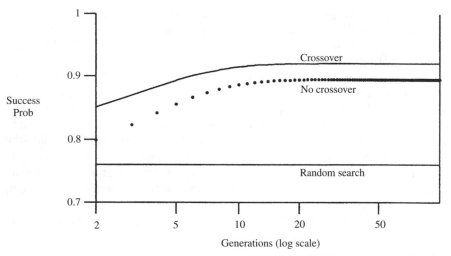

Figure 4-8 The probability of the population containing the global best solution as a function of the number of generations for the case of {0, 1, 2, 3} being represented by {00, 01, 10, 11}, respectively. Crossover is seen to always offer a higher probability of success than relying on mutation alone, or a completely random search (from De Jong et al., 1995).

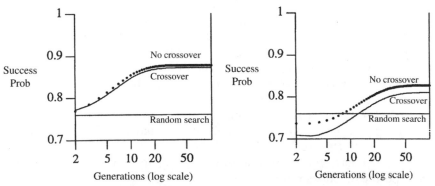

Figure 4-9 The probability of the population containing the global best solution as a function of the number of generations for the cases of the two other equivalence classes depending on the permutation of representing {0, 1, 2, 3} by elements of {00, 01, 10, 11}. Crossover is seen to always offer a lower probability of success than relying on mutation alone, and in the third class the completely random search outperforms evolutionary search for the first 8–12 generations (from De Jong et al., 1995).

lating candidate solutions to the problem at hand. The no free lunch theorem establishes that there is no best evolutionary algorithm across all problems. What about the choice of representation? Recall that in the mid-1970s there was speculation that binary representations would be optimal because they offer the greatest intrinsic parallelism (Holland, 1975, p. 71). By the no free lunch theorem, we know that this cannot be correct across all problems: Intrinsic parallelism offers no benefit across all problems. But what about a single problem? Here, too, it can be shown that for any evolutionary algorithm using a bijective representation (i.e., a one-to-one and onto mapping from the state space of possible solutions to the state space of encoded solutions), an exactly identical algorithm exists with every other bijective representation (Fogel and Ghozeil, 1997b). That is, anything that can be evolved in base two can be identically evolved, generation for generation and solution for solution, with representations in octal, hex, or base 13 when using the appropriate variation operator(s). Again, intrinsic parallelism provably offers no benefit in terms of "information" gained. By consequence, a suitable recommendation is to use a representation that follows naturally from the problem at hand, where the structure of the problem can be seen intuitively and specific operators can be matched to the representation to take advantage of that structure.

4.3   EMPIRICAL ANALYSIS

4.3.1   Variations of Crossover

Theoretical results regarding the rate of convergence with evolutionary algorithms on various problems have been difficult to obtain. Consequently, many efforts have been made to quantify the performance of various procedures empirically (Grefenstette, 1986; Schaffer et al., 1989; and others). Syswerda (1989) examined the utility of various crossover operators, concentrating on one-point, two-point, and uniform crossover. One-point crossover is as defined on page 105. Two-point crossover operates on two parents, but as the name suggests, two points are selected at random rather than a single point and the sequence of components between the points is exchanged (Figure 4-10a). Uniform crossover generates two offspring from two parents by selecting each component from either parent with a given probability (which was set at 0.5 in Syswerda, 1989) (Figure 4-10b).

Several binary function optimization problems were studied. Of these, three specific functions were

1. $f(\mathbf{x}) = \sum_{i=1}^{30} x_i, \; n = 30$ (4-103)

2. $f(\mathbf{x}) = \sum_{i=1}^{30} x_i, \; n = 300$ (bits 31–300 are irrelevant) (4-104)

(a)
Two-Point Crossover

Parent #1: $x_{1,1} \ x_{1,2} \dots x_{1,c_1} \ x_{1,c_1+1} \dots x_{1,c_2} \ x_{1,c_2+1} \dots x_{1,k}$

Parent #2: $x_{2,1} \ x_{2,2} \dots x_{2,c_1} \ x_{2,c_1+1} \dots x_{2,c_2} \ x_{2,c_2+1} \dots x_{2,k}$

Offspring #1: $x_{1,1} \ x_{1,2} \dots x_{2,c_1} \ x_{2,c_1+1} \dots x_{2,c_2} \ x_{1,c_2+1} \dots x_{1,k}$

Offspring #2: $x_{2,1} \ x_{2,2} \dots x_{1,c_1} \ x_{1,c_1+1} \dots x_{1,c_2} \ x_{2,c_2+1} \dots x_{2,k}$

(b)
Uniform Crossover

Parent #1: $x_{1,1} \ x_{1,2} \dots x_{1,k-1} \ x_{1,k}$

Parent #2: $x_{2,1} \ x_{2,2} \dots x_{2,k-1} \ x_{2,k}$

Offspring #1: $x_{1,1} \ x_{2,2} \dots x_{1,k-1} \ x_{2,k}$

Offspring #2: $x_{2,1} \ x_{1,2} \dots x_{2,k-1} \ x_{1,k}$

Figure 4-10 (a) Two-point crossover operates in a similar manner as one-point crossover. Two points are selected along the coding string (c_1 and c_2), and the segments in between these points are exchanged. (b) Uniform crossover creates two offspring by choosing each component from either parent with a specific probability. The figure depicts one such possible crossover.

3. $\quad f(\mathbf{x}) = \left(K^{-1} + \sum_{j=1}^{25} f_j(\mathbf{x})^{-1} \right)^{-1},$ \hfill (4-105)

$$f_j(\mathbf{x}) = c_j + \sum_{i=1}^{2} \left(x_i - a_{ij} \right)^6,$$

$$[a_{ij}] = \begin{bmatrix} -32 & -16 & 0 & 16 & 32 & -32 & -16 & \dots & 0 & 16 & 32 \\ -32 & -32 & -32 & -32 & -32 & -16 & -16 & \dots & 32 & 32 & 32 \end{bmatrix}$$

$$c_j = j,$$

$$K = 500.$$

The third function is known as Shekel's foxholes, and the specific constants are taken from De Jong (1975). A fourth function involved a 16-city traveling salesman problem in which the cities were uniformly spaced around a circle of radius 45 units. The objective function was chosen as the length of the tour subtracted from 1600, with the goal therefore of maximizing the objective value. The representation assigned a group of six bits to each city, with each group treated as a binary number. These values were sorted into increasing order, and this implied the order of the cities to visit. Fifty trials were conducted on each function and the population size was set at 50, except in the traveling salesman problem for which it was set at 100.

The results for the first two functions are shown in Figures 4-11 and 4-12. Uniform crossover is seen to outperform both two-point and one-point crossover on average, and considerably so in the second problem. Figure 4-13a indicates the mean results for Shekel's foxholes. Two-point crossover performs best, followed by one-point and uniform crossover. Syswerda tried a second set of 50 trials in which the bit positions were randomly shuffled. Figure 4-13b indicates the results for the revised problem. Uniform crossover achieves similar performance as before, but the performance of both one-point and two-point crossover is considerably poorer. Finally, Figure 4-14 indicates the results for the traveling salesman problem. Again, uniform crossover is seen to outperform two-point and one-point crossover. The results for the other problems studied in Syswerda (1989) are similar. Generally, uniform crossover yielded better performance than two-point crossover, which in turn yielded better performance than one-point crossover.

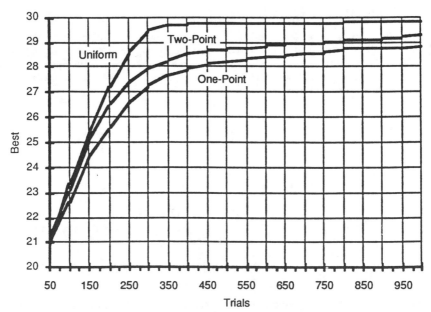

Figure 4-11 Results of one-point, two-point, and uniform crossover on the one max function (after Syswerda, 1989). Uniform crossover provides the best performance.

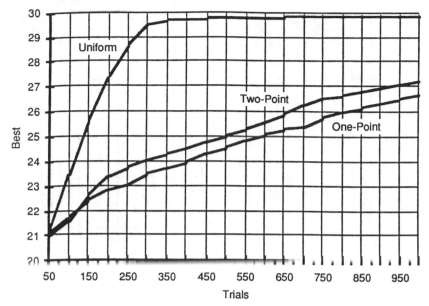

Figure 4-12 Results of one-point, two-point, and uniform crossover on the sparse one max function (after Syswerda, 1989). Uniform crossover provides substantially improved performance.

These results were somewhat surprising (Spears and De Jong, 1991a). The traditional view was that one-point and two-point crossover would be better at assembling and maintaining building blocks (subsections of schema) with above-average performance because these operators are less disruptive than uniform crossover. But Spears and De Jong (1991b) indicated that disruptiveness need not always be viewed negatively. For example, if a population is prematurely converged at a suboptimum point (with a nearly homogeneous population), disrupting the coding strings is the only mechanism for advancing the search.

Uniform crossover is capable of generating more diverse new offspring than is one-point or two-point crossover. Consider the two 10-bit parent strings:

$$P_1: 1\ 1\ 1\ 1\ 1\ 1\ 1\ 1\ 1\ 1$$
$$P_2: 0\ 0\ 0\ 0\ 0\ 0\ 0\ 0\ 0\ 0.$$

Uniform crossover can generate any point in the entire sample space from these two parents, whereas operators relying on a fixed number of cross-over points cannot. But generating diverse samples does not ensure an effective search. Drawing new strings completely at random guarantees the potential for diversity but results in an essentially enumerative procedure. A crucial balance needs

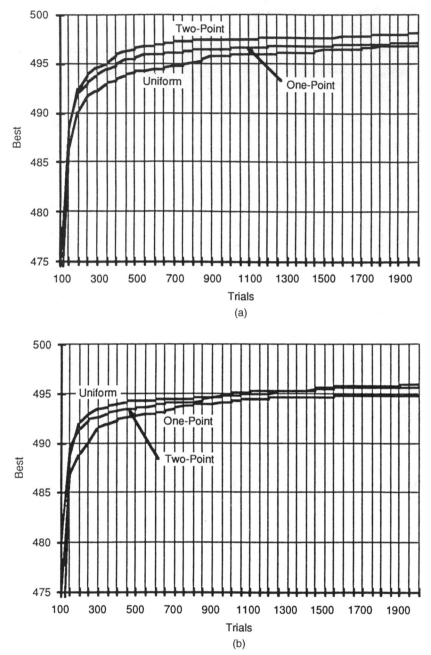

Figure 4-13 (a) Results of one-point, two-point, and uniform crossover on Shekel's foxholes. Two-point crossover is observed to give the best performance. (b) Results comparing the same operators on Shekel's foxholes when the components of the coding strings are randomly shuffled. Uniform crossover achieves similar performance as observed in (a), but two-point and one-point crossover are less effective (after Syswerda, 1989).

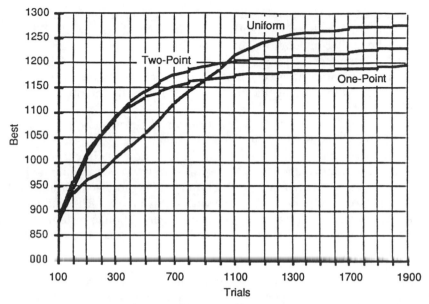

Figure 4-14 Results of one-point, two-point, and uniform crossover on a 16-city traveling sales-man problem. The initial rate of optimization is slower under uniform crossover, but it eventually provides superior performance (after Syswerda, 1989).

to be maintained between *exploration,* generating diverse new trials, and *exploitation,* making use of gains that have already been achieved.

Recall that there is no single optimum operator for searching all possible response surfaces (Wolpert and Macready, 1997). Uniform crossover will outperform *n*-point crossover in some cases, but it is interesting to note that other factors (e.g., population size) can affect the outcome of such comparisons (Spears and De Jong, 1991a). The experiments of Syswerda (1989) and others do, however, indicate the potential for taking a positive view of disrupting schemata. The usefulness of such disruptions will be highly problem dependent, and mechanisms that allow for the self-adaptation (see Section 4.3.5) of search operators (Schwefel, 1981; Fogel et al., 1992) may provide for increased robustness in many settings (Schaffer and Morishima, 1987; Davis, 1989; Spears and De Jong, 1991a, b; Bäck, 1992).

4.3.2 Dynamic Parameter Encoding

Schraudolph and Belew (1992) recognized a potential limitation to using binary encodings for continuous function optimization (as was often done when trying to maximize implicit parallelism). The typical procedure codes real-valued parameters as a series of bits (say, of length *k*), and the choice for the number of bits determines the precision of the search, as well as the size of the sample space. "Unfortunately, the [evolutionary algorithm] searches long place-value codes rather inefficiently" (Schraudolph and Belew, 1992) because much of the early computational effort is directed at the least significant digits of each parameter. The search may prematurely converge on these digits, fixating the population at these values. This leaves a constrained search for the best settings of the most significant digits and is exactly the obverse of what would be desired.

To overcome this difficulty, Schraudolph and Belew (1992) introduced a method for continuously rescaling the precision of each k-bit coded parameter, termed *dynamic parameter encoding* (DPE). Essentially, each parameter of a continuous fitness function, f, is presumed to vary over an interval $[a_i, b_i)$, where i refers to the ith parameter. The binary coding for each parameter then corresponds to some point (or region) within this interval. The interval is divided into subintervals (Figure 4-15). At each generation, a histogram of the population is constructed over the quarters of the current search interval formed by the two most significant bits of the coding for each parameter. The population count for each of three overlapping target intervals is then obtained from the histogram. If the largest count exceeds a preselected threshold (e.g., 90 percent of the population), the algorithm is considered converged and the search is rescaled (*zoomed*) to continue within that subinterval.

Schraudolph and Belew (1992) tested this procedure on the response surfaces examined by De Jong (1975). Table 4-1 indicates the number of parameters, the binary coding length for each parameter, and the functions studied. Schaffer et al. (1989) offered optimal control settings for this test suite. Schraudolph and Belew used these settings (population size of 30, crossover rate = 0.95, mutation rate = 0.005) to execute the GAucsd/GENESIS software package (a particular evolutionary algorithm) with and without DPE. The search resolution for DPE was held constant at $k = 3$ to facilitate comparisons across the test suite.

Figure 4-16 shows the mean results of 10 trials conducted to 10,080 offspring on each function, both with and without DPE (Schraudolph and Belew, 1992). The curves are exponentially smoothed with a window of 10 generations.

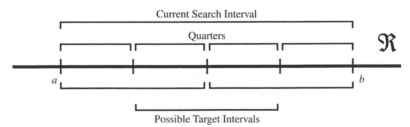

Figure 4-15 Under dynamic parameter encoding, a current search interval (as depicted ranging from a to b) is divided up into possible target intervals. Whenever the histogram of the population appears converged in a particular target interval, the search is rescaled to continue within that interval (after Schraudolph and Belew, 1992).

Table 4-1 The Number of Parameters, The Binary Coding Length, and The Functions Studied in Schraudolph and Belew (1992). These Follow Previous Efforts By De Jong (1975). The Operation $[x_i]$ in $f3$ Returns The Greatest Integer Less Than or Equal To x_i. The $N(0,1)$ in $f4$ Represents A Standard Gaussian Random Variable.

Function	Dimension	Bit Length
$f1$	3	10
$f2$	2	12
$f3$	5	10
$f4$	30	8
$f5$	2	17

Function	Parameter Range
$f1: F(x) = \sum\limits_{i=1}^{3} x_i^2$	$[-5.12, 5.12]$
$f2: F(x) = 100(x_1^2 - x_2)^2 + (1 - x_1)^2$	$[-2.048, 2.048]$
$f3: F(x) = \sum\limits_{i=1}^{5} [x_i],$	$[-5.12, 5.12]$
$f4: F(x) = \sum\limits_{i=1}^{30} i x_i^4 + N(0,1)$	$[-1.28, 1.28]$
$f5: F(x)^{-1} = 1/K + \sum\limits_{j=1}^{25} f_j(x)^{-1},$	$[-65.536, 65.536]$

$$f_j(x) = c_j + \sum_{i=1}^{2} (x_i - a_{ij})^6,$$

where $K = 500$, $c_j = j$, and

$$[a_{ij}] = \begin{bmatrix} -32, -16, & 0, & 16, & 32, -32, -16, \ldots, & 0, 16, 32 \\ -32, -32, & -32, & -32, & -32, -16, -16, \ldots, & 32, 32, 32 \end{bmatrix}$$

Improved performance under DPE is most evident on functions $f1$ and $f3$: The average function values achieved with DPE are 6×10^{-10} and 4×10^{-7} (off the chart), respectively, whereas the standard genetic algorithm stalls out around 10^0 in both cases. Substantial improvements are also evidenced on functions $f2$ and $f4$, but results on $f5$ are actually worse using DPE. Schraudolph and Belew (1992) speculated that this performance was due to the limited number of bits per parameter (i.e., three), and they noted improved performance on $f5$ when the number of bits per parameter was doubled to six.

DPE remains a promising heuristic for improving optimization using evolutionary algorithms that rely on binary (or other low-cardinality) encodings. It is essentially "best characterized as a succession of [evolutionary algorithms] over search spaces iteratively refined through bisection" (Schraudolph

$\bar{f}$ (smoothed)

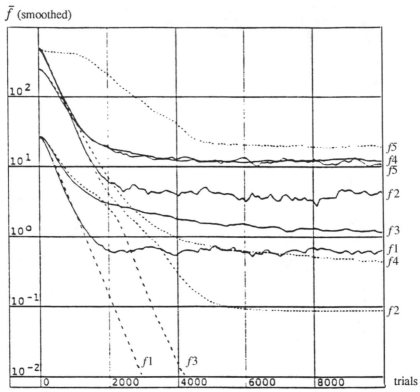

Figure 4-16 The mean results of 10 trials conducted to 10,080 offspring on each function, both with and without dynamic parameter encoding. The solid lines indicate the performance of a standard genetic algorithm, while the dotted lines indicate the performance of dynamic parameter encoding. The curves are exponentially smoothed. Dynamic parameter encoding offers the greatest advantages on functions $f1$ and $f3$, but results on $f5$ are worse than simply using a standard genetic algorithm (after Schraudolph and Belew, 1992).

and Belew, 1992). The zoom operation can only reduce the interval for each parameter as the population converges; no mechanism is offered for expanding the search interval should the global optimum not be contained within the initial specified range in each parameter. Schraudolph and Belew (1992) discussed some of the problems associated with determining the appropriate conditions for expanding these intervals, and they concluded that establishing a well-founded criterion for employing an "inverse zoom" is impossible. Yet it would appear necessary to provide some mechanism for extending the search beyond *a priori* limits. Even randomly perturbing the interval width at each generation might be useful in this regard.

4.3.3   Comparing Crossover to Mutation

Considerable attention has been devoted to assessing the relative importance of crossover and mutation in evolutionary optimization. Holland (1975) observed that mutation and selection is "little more than an enumerative plan retaining the best structure encountered to each point in time." He suggested that mutation's primary role in adaptation is not one of generating new structures ("a role very efficiently filled by crossing-over," Holland, 1975, p. 110). Rather, it ensures that recombination operators have a full range of possible allelic values to explore. Similar comments relating the importance of crossover as opposed to random mutation were offered in Grefenstette (1986), Goldberg (1989, p. 106), Davis (1991, pp. 17–18), Holland (1992), and many others. Koza (1992) went so far as to claim that "Holland's view of the crucial importance of recombination and the relative unimportance of mutation contrasts sharply with the popular misconception of the role of mutation in evolution in nature and with the recurrent efforts to solve adaptive systems problems by merely mutating and saving the best" (p. 27).

Fogel and Atmar (1990) attempted to quantify the benefits, if any, of using genetic operators such as crossover and inversion as opposed to simple random mutation in a series of experiments aimed at solving linear systems of equations. Consider a system of linear equations of n dimensions:

$$b_i = \sum_{j=1}^{n} a_{ij}x_j, i = 1, \ldots, n, \tag{4-106}$$

where the vector $\mathbf{x}$ represents the coding structure of n "gene" products, the vector $\mathbf{b}$ represents n phenotypic behavioral responses, and the coefficients a_{ij} of an $n \times n$ matrix A represent the respective contributions of each component of $\mathbf{x}$ to each component of the response vector $\mathbf{b}$. Any such system will be pleiotropic (single genes expressing themselves through multiple effects) unless $a_{ij} = 0$ for all $i \neq j$.

The fitness function used to evaluate the quality of the evolved behavioral responses to the required response vector was

$$E = \sum_{i=1}^{n} |E_i|, \tag{4-107}$$

where

$$E_i = \sum_{j=1}^{n} a_{ij}x_j - b_i, i = 1, \ldots, n \tag{4-108}$$

Unless the matrix $A = [a_{ij}]$ is singular, there exists an ideal behavioral response $(A\mathbf{x}^*)$. E, the total amount of behavioral error, will equal zero when $x_j = x_j^*$, for $j = 1, \ldots, n$.

Experiments were conducted with an arbitrary coefficient matrix A of rank 10 and a randomly chosen desired response vector $\mathbf{b}$. The population was initialized with 150 vectors (genotypes) taken at random with components distributed normally with zero mean and a standard deviation of 30. These vectors were randomly assigned to be one of three types. The first was altered only by mutation: A standard normally distributed random variable was added to each of its components. The second was altered by one-point crossover (80 percent chance per offspring) and inversion (50 percent chance per offspring). The third was altered by crossover and inversion and also given a 1 percent chance per offspring of random mutation by altering components by a standard Gaussian (as in type 1).

Vectors were scored with respect to the above criterion and were probabilistically selected to survive into the subsequent generation based on a stochastic tournament. Each vector competed against 10 other vectors, with the probability of attaining a "win" equal to the opponent's error score divided by the sum of both vectors' error scores. Once this competition had occurred for all vectors, those with the most wins became the basis set of the next generation.

Linear systems of equations provided a convenient mechanism for examining the effects of these operators in domains of varying interactivity. Five sets of trials were conducted. Each set varied the degree of interactivity of the A matrix by setting the probability of an off-diagonal entry being nonzero to 0.0, 0.25, 0.5, 0.75, and 1.0, respectively. When the probability was zero, the domain was minimally pleiotropic; each independent gene product contributed only to the fitness of its respective component. When the probability was 1.0, the domain was fully pleiotropic, with each gene product contributing to the total behavioral error summed over all behavioral responses. Each experiment consisted of 100 randomly chosen systems. In each system, evolution was halted after 5000 offspring had been evaluated. The results appear in Table 4-2.

Advantage quickly accrued to those genotypes that were altered by random mutation alone. The observed number of trials in which simple random mutation dominated the population is significantly greater than would be expected under a null hypothesis of the behavior of the evolutionary process being independent of the utilized mutation or genetic operation ($P < 0.0001$, nonparametric chi-square test). Furthermore, insufficient evidence exists to suggest that the degree of interactivity of the linear system of equations is associated with the degree to which those genotypes undergoing random mutation alone will overtake the population ($P > 0.12$). In light of these results, further experiments were conducted on independent trials using different rates for crossover and mutation (inversion was omitted), and no statistically significant advantage for crossover was observed.

Fogel and Atmar (1990) concluded that "while specific circumstances (other than linear equations) may well exist for which crossover and inversion operations are especially appropriate, those conditions cannot be the hallmark of a broadly useful algorithm. . . . Rather than being fundamentally different

Table 4-2 The Percentage of Trials When More Than 50 Percent of the Population Consisted of the Given Type (Mutation Alone, Crossover, and Inversion, or Crossover, Inversion, and Mutation). The Values in Parentheses Indicate the Percentage of Trials in Which the Given Types had Completely Taken Over 100 Percent of the Population (after Fogel and Atmar, 1990).

Degree of Interactivity		Percentage of Trials When More Than 50% of the Population Consisted of the Given Type
0	Random Mutation Alone	78% (70%)[a]
	Crossover/Inversion	9% (5%)
	Crossover/Inversion/Mutation	13% (4%)
25	Random Mutation Alone	75% (72%)
	Crossover/Inversion	10% (3%)
	Crossover/Inversion/Mutation	15% (3%)
50	Random Mutation Alone	83% (83%)
	Crossover/Inversion	9% (7%)
	Crossover/Inversion/Mutation	8% (4%)
75	Random Mutation Alone	88% (86%)
	Crossover/Inversion	2% (0%)
	Crossover/Inversion/Mutation	10% (6%)
100	Random Mutation Alone	76% (73%)
	Crossover/Inversion	10% (6%)
	Crossover/Inversion/Mutation	14% (6%)

[a] The values in parentheses indicate the percentage of trials in which the given type had completely taken over 100% of the population.

from random mutation . . . crossover and inversion are merely a subset of all random mutations. As in all subsets, their applicability will be strongly problem dependent, if advantageous at all."

Schaffer and Eshelman (1991) conducted a series of similar experiments comparing crossover with mutation on binary function optimization problems. The simulations started with a randomized population of 500 binary strings. In every problem, each string possessed 100 bits that encoded the trial solution to the task at hand, along with an additional bit that determined whether or not the string would be subject to crossover. Ten percent of the population was randomly chosen to have the crossover bit turned on. Evolution proceeded as follows:

1. Randomly mate the appropriate crossing individuals among themselves; replace them by their offspring.

2. If the selected mutation rate (see below) was positive, perform mutation on the entire gene pool (including the crossover bit).

3. Score all new solutions.

4. Select the best solutions for inclusion in the subsequent population.

5. Halt if two generations have transpired without any new offspring or
   50,000 offspring have been evaluated.

Twenty trials were conducted with each of five functions. The first function
was termed *onemax:* The fitness of a solution was simply the number of ones
in the string. The second function was termed *plateau:* A value of five was
given for each five-bit segment of "00000." The overall evaluation was calculated as 100 minus the sum of the segment scores, so that the optimum score
was zero and the worst possible score was 100. The third function was termed
trap: A value of five was again given for the sequence "00000," but for other
patterns, a value of 0.5 times the number of ones was awarded. Thus, the problem is "deceptive" (Goldberg, 1987) in that the average fitness of short-length
building blocks (schemata) leads away from the global optimum. The fourth
and fifth problems were simply variants of plateau and trap in which the segments were uniformly spaced throughout the 100 bits. That is, the bits of segment 1 were in positions 1, 21, 41, 61, and 81; the bits of segment 2 were in positions 2, 22, 42, 62, and 82, and so forth.

Mutation was evaluated in two forms: zero mutation and a rate of 0.0005.
The latter rate was chosen empirically so as to try to balance the number of
solutions that improve from both the mutation only and crossover subpopula-
tions in the initial generation (Schaffer and Eshelman, 1991).

The experiments were first executed 20 times each with mutation alone,
two-point crossover alone, and uniform crossover alone (i.e., no mutation) in
order to provide a baseline for comparison. The population size was set at 500
in all cases. Subsequent experiments examined the tendency for solutions undergoing either two-point or uniform crossover to invade the population, with
or without mutation.

The results were mixed. Four separate characteristics were observed:
(1) the solutions created by crossover dominated the population and achieved
a level of performance not attained by mutation alone, (2) the solutions created by crossover dominated the population and achieved a level of performance that was about the same as mutation alone but was achieved more
quickly, (3) the solutions created by crossover dominated the population but
the performance was ultimately worse than achieved by mutation alone, and
(4) the solutions by crossover failed to dominate the population and performance was worse than achieved by mutation alone. When combined with mutation, both methods of crossover always came to dominate the population.

Results that were superior to mutation alone were achieved by two-point crossover on the trap function, with and without mutation. But results that were inferior to mutation alone were achieved by uniform crossover on the plateau function, with and without mutation, and other combinations of two-point and uniform crossover on the various versions of plateau and trap also generated inferior results.

Broadly, the results again indicated that the utility of specific crossover and mutation operations is problem dependent (as is now known to be true by the no free lunch theorem). As Schaffer and Eshelman (1991) stated:

> We believe Fogel and Atmar are correct to point out the power of selection and mutation for search. This power is often underestimated in the [genetic algorithm] community. We believe it is not correct to infer from their results that crossover provides no added value. We believe that our experimental results begin to shed more light on the circumstances under which crossover will deliver the search power first described by Holland. However, these results also suggest circumstances in which crossover has a detrimental impact on search. It must be used with care.

Schaffer and Eshelman (1991) speculated that the results favoring mutation reported in Fogel and Atmar (1990) may have been due to: (1) coupling crossover with an inversion operator that actually reordered the coding string, rather than simply reindexing its components, and (2) applying crossover to real-valued instead of binary coding, preventing the possibility for crossover to perform any within-parameter search.

In light of these criticisms, Fogel and Stayton (1994) compared a simple evolutionary algorithm relying solely on real-valued coding and Gaussian mutation with the results offered in Schraudolph and Belew (1992), both with and without dynamic parameter encoding. Schraudolph (personal communication) kindly provided the raw data from their experiments, which as described above relied on binary coding and achieved some of the best optimization results on the test suite offered in De Jong (1975).

The simple evolutionary algorithm was conducted as follows:

1. Thirty real-valued vectors, x_i, $i = 1, \ldots, 30$, were initialized by sampling from a uniform distribution across the preselected range of each parameter (De Jong, 1975).

2. All vectors, x_i, $i = 1, \ldots, 30$, were scored with respect to the chosen function, $F(x)$.

3. From each vector, x_i, $i = 1, \ldots, 30$, one offspring was created, denoted by x_{i+P}, by adding a Gaussian random variable with zero mean and variance equal to $F(x_i)/n^2$ to each component, where $F(x_i)$ represents the parent's error score and there are n dimensions in the optimization problem.

4. Each offspring, $\mathbf{x}_i$, $i = P + 1, \ldots, 2P$, was scored with respect to $F(\mathbf{x})$.

5. For each $\mathbf{x}_i$, $i = 1, \ldots, 2P$, 10 competitors were selected at random from the population. Pairwise comparisons were made between $\mathbf{x}_i$ and each of the competitors. If the error score of $\mathbf{x}_i$ was less than or equal to that of its opponent, it was assigned a win.

6. The 30 vectors with the greatest number of wins were selected to be parents for the next generation.

7. The process repeated to step 3 until 10,080 vectors (336 generations) were evaluated.

Although the population was initialized randomly over the parameter ranges noted in Table 4-1, solutions were allowed to vary outside these ranges in subsequent generations. Unlike the experiments with dynamic parameter encoding, the global minimum was not required to be bounded by the initial search limits. One exception to this was required for the step function ($f3$), which continues stepping to successively lower values as the parameters tend toward negative infinity. For this problem, if any parameter of a new trial solution exceeded its initial limits, it was set equal to the limit it exceeded. For the noisy quartic function ($f4$), all trial solutions were reevaluated every generation. In the event that a solution's score was negative, the standard deviation for generating new trials from this solution was set equal to the small positive constant 10^{-3}.

Of the five test functions from De Jong (1975), only the fifth possesses multiple local optima. This function is somewhat pathological. To provide additional tests with more reasonable surfaces that possess local optima, three two-dimensional functions (Figure 4-17) were taken from Bohachevsky et al. (1986):

$$f6: F(x, y) = x^2 + 2y^2 - 0.3\cos(3\pi x) - 0.4\cos(4\pi y) + 0.7,$$

$$f7: F(x, y) = x^2 + 2y^2 - 0.3(\cos(3\pi x)\cos(4\pi y)) + 0.3,$$

$$f8: F(x, y) = x^2 + 2y^2 - 0.3(\cos(3\pi x) + \cos(4\pi y)) + 0.3.$$

Schraudolph (personal communication) provided the GAucsd/GENESIS software, and results were obtained on $f6$–$f8$ with and without DPE using the same settings as offered in Schraudolph and Belew (1992). Ten trials were conducted with and without DPE on each function. For standard experiments without DPE, the available range for each parameter was $[-50, 50]$ with a coding length of 14 bits. Experiments with DPE used three bits following Schraudolph and Belew (1992). Five hundred trials were conducted with the simple evolutionary algorithm on each function $f1$–$f8$.

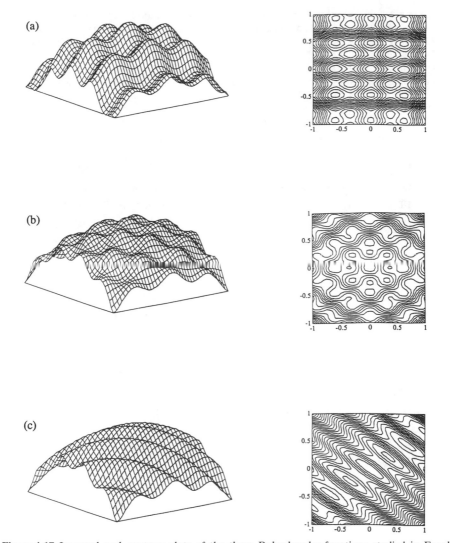

Figure 4-17 Inverted and contour plots of the three Bohachevsky functions studied in Fogel and Stayton (1994). (a) $F(x,y) = x^2 + 2y^2 - 0.3\cos(3\pi x) - 0.4\cos(4\pi y) + 0.7$. (b) $F(x,y) = x^2 + 2y^2 - 0.3(\cos(3\pi x)\cos(4\pi y)) + 0.3$. (c) $F(x,y) = x^2 + 2y^2 - 0.3(\cos(3\pi x) + \cos(4\pi y)) + 0.3$.

Table 4-3 indicates the best score in the population and the mean of all parents' scores after 10,080 function evaluations. The simple evolutionary algorithm outperformed the algorithm relying heavily on crossover, both with and without DPE on functions $f1$, $f2$, and $f6$–$f8$, and generated comparable

Table 4-3 Results for the Best Score in the Population and the Mean of All Parents' Scores After 10,080 Function Evaluations, Averaged Over 10 Trials With the Genetic Algorithm Techniques (Both With And Without Dynamic Parameter Encoding) and 500 Trials With "Evolutionary Programming" (after Fogel and Stayton, 1994). Evolutionary Programming Outperforms Both Genetic Methods On Functions $f1$, $f2$, and $f6$–$f8$, and Yields Comparable Performance on $f3$–$f5$. The Values in Parentheses Indicate the Standard Deviations.

		Average Best	Average Mean
$f1$:	EP	3.149×10^{-66} (2.344×10^{-129})	1.087×10^{-65} (1.794×10^{-128})
	DPE	1.056×10^{-11} (1.072×10^{-21})	3.618×10^{-10} (1.060×10^{-18})
	GA	2.836×10^{-4} (4.587×10^{-8})	6.135×10^{-1} (1.627×10^{-1})
$f2$:	EP	1.215×10^{-14} (3.357×10^{-26})	8.880×10^{-14} (2.399×10^{-24})
	DPE	2.035×10^{-2} (3.315×10^{-3})	8.785×10^{-2} (8.090×10^{-3})
	GA	2.914×10^{-2} (6.280×10^{-4})	1.722×10^{0} (2.576×10^{0})
$f3$:	EP	0.0 (0.0)	0.0 (0.0)
	DPE	0.0 (0.0)	0.0 (0.0)
	GA	0.0 (0.0)	1.307×10^{0} (1.172×10^{-1})
$f4$:	EP	-2.575×10^{0} (7.880×10^{-1})	-4.274×10^{-1} (3.206×10^{-2})
	DPE	-2.980×10^{0} (1.009×10^{-1})	4.704×10^{-1} (1.283×10^{-1})
	GA	-4.599×10^{-1} (4.265×10^{-1})	1.312×10^{1} (2.941×10^{0})
$f5$:	EP	4.168×10^{0} (9.928×10^{0})	4.194×10^{0} (1.022×10^{1})
	DPE	3.502×10^{0} (1.265×10^{1})	1.642×10^{1} (7.101×10^{2})
	GA	9.980×10^{-1} (3.553×10^{-15})	1.021×10^{1} (7.165×10^{1})
$f6$:	EP	5.193×10^{-96} (1.348×10^{-188})	9.392×10^{-94} (4.410×10^{-184})
	DPE	1.479×10^{-9} (1.460×10^{-18})	8.340×10^{-7} (4.649×10^{-14})
	GA	2.629×10^{-3} (1.013×10^{-5})	4.022×10^{1} (6.467×10^{3})
$f7$:	EP	8.332×10^{-101} (3.449×10^{-198})	2.495×10^{-99} (3.095×10^{-195})
	DPE	2.084×10^{-9} (6.831×10^{-18})	6.520×10^{-7} (7.868×10^{-14})
	GA	4.781×10^{-3} (2.146×10^{-5})	3.541×10^{1} (2.922×10^{3})
$f8$:	EP	1.366×10^{-105} (4.479×10^{-208})	3.031×10^{-103} (2.122×10^{-203})
	DPE	1.215×10^{-5} (5.176×10^{-10})	3.764×10^{-1} (9.145×10^{-1})
	GA	2.444×10^{-3} (4.511×10^{-5})	2.788×10^{-1} (1.368×10^{-3})

performance on $f3$–$f5$. Although the statistical significance of the data was often great (e.g., $P < 10^{-6}$ when comparing the mean best discovered solution using mutation alone to the GAucsd/GENESIS on $f1$), in many cases the data did not provide statistically significant evidence for a difference in the expected outcome when using dynamic parameter encoding, despite a difference in performance of many orders of magnitude. This was due mainly to the small sample size provided in Schraudolph and Belew (1992) and the relatively large variance associated with their results.

The evidence (and the underlying mathematics) consistently suggests that there is no requirement for incorporating recombination in evolutionary algorithms, nor does its inclusion always generate improvements over searches relying solely on mutation (cf. Davis, 1991, p. 18). The representation used to describe alternative solutions in part determines the shape of the response surface being searched; thus, the utility of various transformation operations will be dependent on representation as well. But no consistent advantage accrues from representing real-valued parameters as binary strings and allowing crossover to do within-parameter search (cf. Schaffer and Eshelman, 1991). Evolutionary algorithms that rely solely on mutation can, in many cases, be used to greater advantage on function optimization problems.

1.3.1 Crossover as a Macromutation

The traditional view of recombination, in particular crossover, has been that the operator is successful when there are building blocks of independent solutions that can be brought together advantageously. One of the typical methods that was used in the 1980s to determine whether or not crossover was beneficial to an evolutionary algorithm was to test that algorithm with and without crossover. If the results were better with crossover, then this was viewed as evidence in favor of the existence of building blocks. Unfortunately, this line of reasoning is flawed: Just because crossover may be beneficial in a particular situation does not by consequence imply anything about the existence of building blocks.

Jones (1995) was the first to demonstrate this effect clearly within the following context. Consider two alternative search operators: The first performs a typical crossover between two parents in an existing population, and the second performs a crossover between an existing parent and a completely randomly generated new solution. In the latter case, the "random crossover," the operation is purely mechanical: There is no concept of transferring building blocks that are associated with above-average performance when splicing up randomly generated solutions. Jones (1995) wrote cogently: "without the idea of crossover, you do not have crossover. To call this operation crossover is similar to calling a headless chicken running around the yard a chicken. Certainly it has many characteristics of a chicken, but an important piece is missing. A further argument that this operation is not really crossover is that two parents are not required. For example, for two-point crossover on strings, simply choose the crossover points and set all the loci between the points to randomly chosen alleles. This is an identical operation and is clearly just a macromutation."

Jones (1995) reported trials on a variety of problems comprising both cases where well-defined building blocks did and did not exist. In the cases where building blocks were present, traditional crossover outperformed random crossover; however, on the other problems, random crossover outperformed standard crossover. Thus, the mechanism of crossover was useful but merely by serving as a large mutation. Any advantage that occurred due to crossover in these cases was occurring in spite of the mechanism of variation rather than because of its ability to promote useful building blocks.

This result supports earlier speculation offered in Fogel (1994b): Crossover can be useful in the beginning generations of a randomly initialized population simply because it provides a variation with a large step size. Moreover, most of the research in evolutionary algorithms during the mid-1980s to early-1990s relied on high rates of crossover and very low rates of mutation. Thus, to rely solely on mutation in this form would hamper exploration. The majority of test functions studied had structure that promoted taking large steps after random initialization. As Fogel (1994b) observed: "Most completely randomly selected solutions are likely to be poor, and large initial steps away from such points typically offer a greater chance of discovering improved solutions than would concentrating the search around each solution's local neighborhood." But this situation is easily overcome by using variation operators with a continuous range of probability density (e.g., Gaussian operators, as seen in the previous section) and applying them to all parameters of a candidate solution simultaneously.

The "headless chicken" crossover operator has been tested in a variety of other experiments, including those that rely on evolving symbolic expressions in the form of data trees. In this case, sections of independent trees (i.e., subtrees) may act as functional subroutines and therefore may be usefully recombined. Angeline (1997), however, showed that crossing over extant trees with completely randomly generated trees, as opposed to crossing them with other existing trees in the population, generated superior performance on problems involving classification (determining if a point lies on one of two intertwined spirals) and time series prediction (sunspot data). Crossover served simply as a macromutation. Further independent comparisons of "subtree crossover" and "subtree mutations" on a diverse set of benchmark functions have indicated no general advantage to either method (Chellapilla, 1997, 1998; Luke and Spector, 1998; Fuchs, 1998; and others). The available evidence suggests consistently that no single operator (i.e., no specific crossover, mutation, or other procedure) can best treat even the limited set of test problems that are commonly addressed in the literature, let alone provide a more general capability across a wider domain of real-world applications.

4.3.5 Self-Adaptation in Evolutionary Algorithms

Reed et al. (1967), Rosenberg (1967), and Rechenberg (personal communication) all independently introduced the possibility of allowing an evolutionary algorithm to adapt the distribution of new trials in light of information gained during the search in 1967. Following Rechenberg's efforts (see Schwefel, 1981), the application of independent, identically distributed (i.i.d.) Gaussian random perturbations to each component of every solution distributes new trials in alignment with the coordinate axes (Figure 4-18). But the optimum search direction (e.g., the gradient) is rarely so oriented. In general, the search is retarded as the population zig-zags across the gradient. Incorporating correlated mutations across the components of each parent vector provides a mechanism for significantly accelerating the rate of optimization.

As noted above, Schwefel (1981) described a procedure wherein each solution contains an n-dimensional vector of object variables $\mathbf{x}$ and as many as $n(n + 1)/2$ additional mutation parameters, including standard deviations and correlations, describing an n-dimensional Gaussian random vector. These values are modified during the search as

$$\sigma_i' = \sigma_i \cdot \exp(\tau_0 \cdot \Delta\sigma_0) \cdot \exp(\tau \cdot \Delta\sigma_i) \tag{4-109}$$

$$\alpha_j' = \alpha_j + \beta \cdot \Delta\alpha_j \tag{4-110}$$

$$x_i' = x_i + z_i(\sigma', \alpha') \tag{4-111}$$

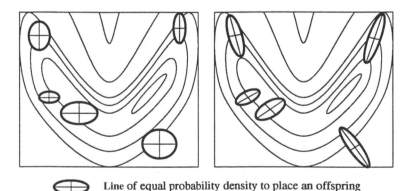

⊕ Line of equal probability density to place an offspring

Figure 4-18 Self-adapting the mutation distributions on each solution allows for new trials to be generated in light of the contours of the response surface. Independent adjustment of the standard deviation in each dimension provides a mechanism for varying the width of the probability contour in alignment with each axis (left picture). Correlated standard deviations provide a mechanism for generating trials such that the probability contours are not aligned with the coordinate axes (right picture) (after Bäck et al., 1991).

where σ is the vector of standard deviations, α is the vector of rotation angles equivalent to the correlations between parameters, and $\mathbf{x}$ is the vector of object values to be manipulated in optimizing the objective function (see page 127 for particular recommended settings of specific parameters). The mutation parameters self-adapt in that both the standard deviations and rotation angles undergo perturbations (lognormal and Gaussian, respectively) based on their own current values. The resulting search can adapt to arbitrary contours on the response surface (Figure 4-18).

Fogel et al. (1991) independently offered a similar procedure for self-adapting uncorrelated standard deviations of Gaussian perturbations. Each solution was coded as the vector of object variables, $\mathbf{x}$, and the associated vector of standard deviations, σ, and offspring were generated according to

$$x_i' = x_i + \sigma_i \cdot z_i \tag{4-112}$$

$$\sigma_i' = \sigma_i + \zeta \sigma_i \cdot z_i \tag{4-113}$$

where x_i and σ_i are the ith components of x and σ, respectively, z_i follows a standard normal distribution, and ζ is a scaling constant chosen to ensure that each σ_i tends to remain positive. Whenever a mutated standard deviation becomes nonpositive, it is set to a small value, $\varepsilon > 0$.

Fogel et al. (1992), at the suggestion of Sebald (personal communication), provided a method for altering correlation coefficients between components. In addition to the self-adaptation of the standard deviations, a correlation coefficient matrix, $p = [\rho_{ij}]$, $i \neq j$, $\rho_{ij} = \rho_{ji}$, is incorporated. The components of p are initialized uniformly over $[-1, 1]$, and mutation of this matrix is conducted by perturbing each component by adding i.i.d. normal random variables with zero mean and preselected standard deviation (e.g., 0.1). If the upper or lower bounds on any entry of p exceed the range $[-1, 1]$, it is reset to the limit it exceeds.

Although both the methods of Schwefel (1981) and Fogel et al. (1991, 1992) are conceptually quite similar, they generate significant differences in performance on selected test functions. Initial comparisons between the two approaches were restricted to uncorrelated self-adaptation. Bäck et al. (1993) evaluated a (30, 200)−EA using the methods of Schwefel and a (200 + 200)−EA using the methods of Fogel et al. on several functions, including the sphere model, a scaled version of the sphere model, and the 30-dimensional multimodal function offered in Ackley (1987, pp. 13–14). The results indicated a general advantage to the methods of Schwefel (1981), although in subsequent discussion with Bäck (personal communication) it was agreed that these results may in some cases reflect the difference in parent population size as well as the choice of scaling factors. Saravanan and Fogel (1994) compared the methods using a (200 + 200)−EA on five different functions including the sphere model, the Rosenbrock function, a

Bohachevsky function, and other functions of 20 dimensions possessing local optima (e.g., generalized Ackley's function). The results more clearly indicated an advantage to the lognormal perturbation scheme offered by Schwefel (1981).

There are three likely reasons for the observed advantage of the lognormal perturbation of the standard deviations over the Gaussian perturbation. First, the lognormal perturbation is guaranteed to yield strictly positive values, alleviating the necessity of selecting a value of ε as is required in the method of Fogel et al. (1991). Second, it is easier for the standard deviations to vary from small values to large values under lognormal perturbations. Under Gaussian perturbations, once the standard deviations become small (e.g., ε), it may take a very low probability mutation to make a jump to a larger value. The search is overconstrained for the test functions studied. Third, the perturbation method of Schwefel (1981) incorporates both a global factor allowing for the overall change of the mutability of all components and local factors for adjusting the individual step sizes. This generates the potential for greater diversity than is allowed in the methods of Fogel et al. (1991, 1992), which again appears to be an advantage for the test functions under consideration. Interestingly, however, Angeline (1996) showed that when noise was added to a similar set of test functions, the Gaussian update strategy appeared to perform better than the lognormal strategy. There is no currently accepted theory that explains or suggests which method to use in different circumstances.

4.3.6 Fitness Distributions of Search Operators

In general terms, the majority of evolutionary optimization algorithms can be described as a series of operators applied to a population of candidate solutions $\mathbf{x}$:

$$\mathbf{x}[t + 1] = s(v(\mathbf{x}[t])) \qquad (4\text{-}114)$$

where $\mathbf{x}[t]$ is the population of solutions at iteration t under a specified representation, $v(.)$ is the random variation operator, and $s(.)$ is the selection operator. This formulation leads directly to a Markov chain view of evolutionary algorithms in which a time-invariant memoryless probability transition matrix describes the likelihood of transitioning to each next possible population configuration given each current possible configuration. The treatment of evolutionary algorithms as Markov chains has already been discussed above. But the description of Eq. 4-114 also suggests that some level of understanding of the behavior of an evolutionary algorithm can be garnered by examining the stochastic effects of the operators $s(.)$ and $v(.)$ on a population $\mathbf{x}$ at time t. Of interest is the probabilistic description of the fitness of the solutions contained in $\mathbf{x}[t + 1]$. Efforts in Altenberg (1995), Grefenstette (1995), Fogel (1995), Nordin and Banzhaf (1995), Fogel and Ghozeil (1996), Kallel and Schoenauer (1997), and others have recently been directed at generalized expressions describing the relationship between offspring and parent fitness under particular variation operators, or empirical determination of the fitness of offspring for a given random variation technique. These "fitness distributions" can be usefully compared to determine if a certain operator should be favored over another in a particular setting.

For example, Fogel and Ghozeil (1996) considered the case of minimizing three different functions:

$$f_1 = \sum_{i=1}^{n} x_i^2 \tag{4-115}$$

$$f_2 = \sum_{i=1}^{n-1} \left(100\left(x_i^2 - x_{i+1}\right)^2 + \left(1 - x_i^2\right) \right) \tag{4-116}$$

$$f_3 = \sum_{i=1}^{n-1} \left(x_i^2 + 2x_{i+1}^2 - 0.3\cos\left(3\pi x_i\right) - 0.4\cos\left(4\pi x_{i+1}\right) + 0.7 \right) \tag{4-117}$$

in $n = 2, 5$, and 10 dimensions. A series of 100 sets of 5000 trials was conducted on each surface for each dimensionality by using a uniform distribution to initialize a population of 100 solutions over $[-4, 4]^n$. In each trial, the best of the initial 100 solutions was mutated to create 100 offspring by adding a multivariate Gaussian random vector with a mean of zero and a single preset standard deviation in each dimension. The standard deviation was incremented from 0.01 to 2.0 stepping by 0.02 (thus accounting for the 100 sets of 5000 trials). The expected improvement was defined as the mean improvement of the best offspring as compared to the (best) parent from the initial population, with the average taken over all trials at a given setting of the standard deviation. Thus, this value could yield a negative result if the best of 100 offspring were worse than the parent that generated it (the procedure was essentially a $(1, 100)-$EA). The probability of improvement was defined as the mean fraction of offspring that outperformed their parent taken across all 100 offspring in each of the 5000 trials at each particular setting of the standard deviation.

As the standard deviation (step size) is increased from 0.01 to 2.0, the probability of improvement and the expected improvement vary. The probability of improvement is maximized at 0.5 on these functions for an infinitesimally small standard deviation. But the expected improvement is essentially zero under these conditions. As the step size is increased, the expected improvement also increases up to the point where the step size is too large for the local basin of attraction, after which it decreases. Thus, it is possible to trade off the probability of improvement for the expected improvement, which offers the potential for finding an optimal setting for the standard deviation given the initial conditions of the population.

As noted in Section 4.2.6, Rechenberg (1973) showed that the maximum expected improvement of a $(1 + 1)-$EA could nearly be obtained by having a probability of success of 0.2 (i.e., the 1/5 rule) for the quadratic and corridor functions as the number of dimensions went to infinity, but these are just two possible functions. Figure 4-19 shows the empirical evidence obtained on f_2 for $n = 2$ and 10. (For space consideration, only f_2 is discussed here.) Recall that each datum represents the mean of 5000 trials at the associated setting of the standard deviation. Note that the standardized expected improvement peaks at a probability of improvement of about 0.2 for the case of 10 dimensions but reaches its maximum at a probability of improvement of about 0.06 in two dimensions. Applying the 1/5 rule in this latter case would result in suboptimal expected improvement. Figure 4-20 shows the log probability of improvement as a function of the standard deviation for $n = 2, 5,$ and 10, while Figure 4-21 shows the standardized expected improvement as a function of the standard

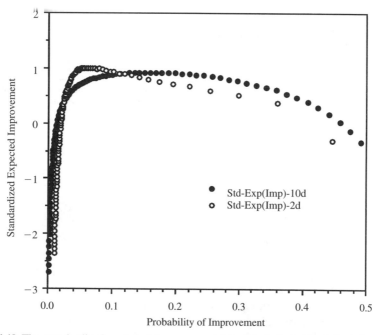

Figure 4-19 The standardized expected improvement (described in standard deviations across the data) as a function of the probability of improvement on the 2-dimensional and 10-dimensional Rosenbrock functions. The maximum improvement for the 10-dimensional case occurs when the probability for improvement is about 0.2. But for the 2-dimensional case it occurs with a probability of improvement of 0.06, and this does not correspond well with the 1/5 rule of Rechenberg (1973) (from Fogel and Ghozeil, 1996).

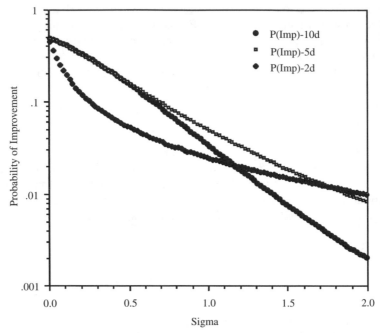

Figure 4-20 The log probability of improvement as a function of the standard deviation on the 2-, 5-, and 10-dimensional versions of the Rosenbrock function (from Fogel and Ghozeil, 1996).

deviation for $n = 2$ and 10. Combining Figures 4-20 and 4-21 yields Figure 4-19.

Statistics on the distribution of fitness scores of offspring can also be useful in determining appropriate variation operators in discrete optimization problems. Fogel and Ghozeil (1996) studied the application of a reversal operator on permutations that represented tours in a traveling salesman problem. For example, the vector [1 2 3 4 5 6] would indicate starting at city #1 and proceeding to each city in the list, then returning to the starting city. The degree of change to a parent tour is dependent in part on the length of the segment being reversed. That is, it is often true that swapping the order of adjacent cities in a tour will have less effect on the tour length than will reversing the order of say, five cities, particularly in a tour that is already well optimized. Fogel and Ghozeil (1996) took the case of a known optimum 30-city tour (see Figure 4-22a) and perturbed it using a series of reversal operators of length 1 to 4, where a reversal of length 1 corresponds to swapping adjacent cities. Figure 4-22b shows a sample perturbed rout-

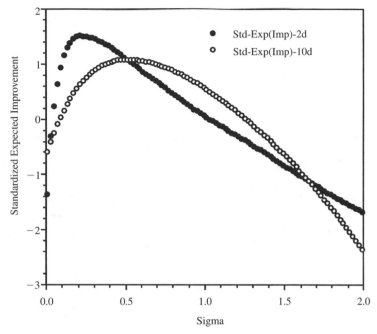

Figure 4-21 The standardized expected improvement as a function of the standard deviation on
the 2- and 10-dimensional versions of the Bohachevsky function. The optimum improvement occurs for different values of the standard deviation (sigma) depending
on the dimensionality of the function (from Fogel and Ghozeil, 1996).

ing. Attention was devoted to determining the differences in the probability
of improvement and expected improvement for all possible reversals of
length 1–15. (Due to the symmetry of traversing the tour clockwise or
counterclockwise, reversals longer than 15 are not required.) Each class of
reversal had 30 potential implementations (e.g., there are 30 different ways
to swap adjacent cities on a 30-city tour), and the results of each possible
instantiation of the operator were enumerated, yielding 900 data points
for each length of reversal taken across all trials. The probability of improvement for a reversal length was calculated as the number of improved tours
generated divided by 900. The expected improvement was calculated as the
ratio:

$$\frac{\left[\sum_{n=1}^{900} \max\left(Old\ Tour\ Length\ -\ New\ Tour\ Length,\ 0\right)\right]}{900}.$$

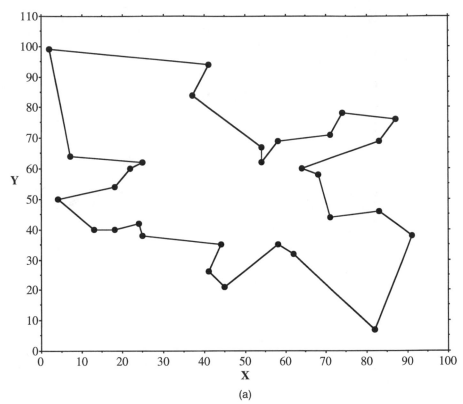

(a)

Figure 4-22 (a) The global best solution to the 30-city traveling salesman problem offered in
Oliver et al. (1987).

Figure 4-23 shows the probability of improvement and mean improvement as a function of the length of the reversal. As the length of the reversal
was increased, both the probability of improvement and the mean improvement decreased. Beyond a reversal of length nine, no improvements were
recorded in any of the 30 cases. Note that the greatest mean improvement occurred for a probability of improvement of close to 0.008, which again does not
correspond well with the 1/5 rule, further illustrating the limited utility of this
heuristic.

This technique can also be used to determine the effectiveness of recombination operators. Nordin and Banzhaf (1995) quantified the fitness change
that occurred after crossover was applied to machine code expressions that
served as regression equations. Figure 4-24 shows the percentage of fitness
change (positive is an improvement) as a histogram over 35 successive generations. Note that the most common effects are either no change at all, mean-

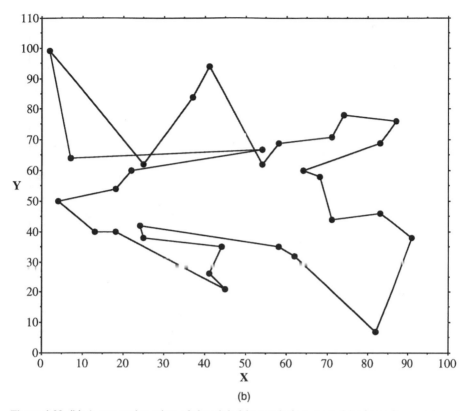

(b)

Figure 4-22 (b) A mutated version of the global best solution created by imposing a series of reversals to the global optimum routing (from Fogel and Ghozeil, 1996).

ing that the offspring's fitness is the same as that of the parents, or a 100 percent or larger decrease in fitness. The construction of superior solutions from "building blocks" of useful code appears to be absent. Moreover, the ineffectuality of crossover increases with increasing generations: More and more, crossover does less and less. This effect probably occurs because crossover mostly recombines sections of different parents that have no effect on the fitness of the solution, and these sections become more frequent as evolution proceeds. Banzhaf et al. (1998) contend that based on these results and other similar results offered in Teller (1996) and Nordin et al. (1996) "one must conclude that traditional . . . crossover acts primarily as a macromutation operator" when evolving expressions that are interpreted as programs (p. 155). This leaves open the opportunity to tailor recombination and other operators to specific problems to take advantage of their structure. Statistics of fitness distributions of operators offer one basis for assessing the effectiveness of different design choices in evolutionary algorithms.

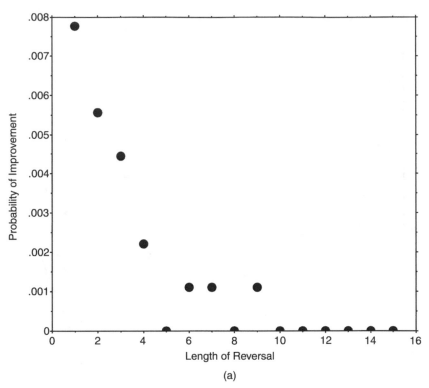

Figure 4-23 (a) The mean probability of improvement.

4.4   DISCUSSION

The most fundamental property of an optimization algorithm is its convergence in the limit. Evolutionary algorithms can be constructed so as to asymptotically converge to globally optimal solutions by properly choosing the variation and selection operators. Whenever variation offers the potential for visiting every possible candidate solution (in a finite set of solutions) and selection retains the best solution discovered at each generation, the evolutionary algorithm will converge with probability 1 to a best solution. In contrast, when using other forms of selection that are not "elitist" (such as roulette wheel selection), or when using variation operators that cannot escape a local neighborhood even when applied over successive generations, global convergence is not guaranteed.

Of more practical interest are the rates of convergence that can be obtained when evolutionary algorithms are applied to specific problems or specific classes of problems. Determining these rates has been difficult except in

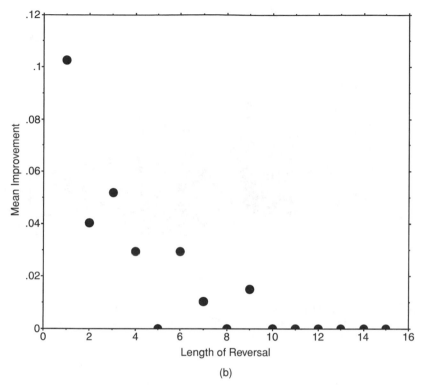

(b)

Figure 4-23 (b) The mean improvement as a function of the length of reversal taken across 30 trials with 30 different perturbed 30-city traveling salesman problems. The best performance is exhibited from the 1-length reversal (from Fogel and Ghozeil, 1996).

the case of strongly convex and other simplified functions. Typical convergence rates of evolutionary algorithms on strongly convex functions are of
geometric order, but this does not compare well with gradient and higher-
order descent methods on these functions. It is unfortunate that most of the
available convergence rate information is applicable only to those functions
where evolutionary algorithms are not the technique of choice. Although
many response surfaces can be approximated by quadratic or linear functions,
using these approximations to determine local convergence rates are only of
limited practical utility.

    The difficulty in obtaining practical convergence rate information has en-
gendered a large volume of empirical research that investigates the application of evolutionary algorithms on specific problems. Different control parameters such as population size, rates of crossover and mutation, stringency of
selection, and so forth, have been compared on benchmark functions. But the
problem of empirically assessing the utility of various optimization methods

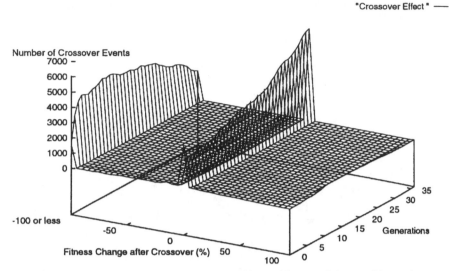

Effects of Crossover during Evolution

"Crossover Effect" ——

Figure 4-24 The fitness change after crossover on the problems evolving machine code expressions from Nordin and Banzhaf (1995). The vast majority of crossovers either result in no change or a large decrease in offspring fitness.

should not be underestimated. As Whitley (1993) remarked, "even if one is interested in optimization the results of any empirical test designed to demonstrate a 'more effective optimizer' may say as much about the nature of the functions being optimized as it does about any two algorithms which might be compared." Indeed, there is no doubt about this: The no free lunch theorem (Wolpert and Macready, 1997) proves that improved performance for any algorithm indicates a match between the structure of the algorithm and the structure of the problem.

The no free lunch theorem ought at least to temper interest in relying solely on empirical comparisons, and although theory is always more informative than anecdotal evidence, it appears that empirical trials still hold the ability to persuade. Whenever a new algorithm is presented as being well suited for a problem, it becomes mandatory to ask why it is well suited and for what problems will it not be appropriate. Sometimes these questions can be answered by further empirical trials, which effectively provide counterexamples. For example, Salomon (1996) examined an evolutionary algorithm that relies heavily on recombination and had demonstrated promising rates of optimization on multimodal functions (e.g., Rastrigin's function). Salomon (1996) showed, however, that the same procedure has a significant degradation in performance when these functions are simply rotated. The structure of these particular problems seemed to be perfectly matched to the evolutionary algorithm only when their dimensions were linearly separable. But as with convex functions, many common techniques can treat linearly separable problems more effectively than evolutionary algorithms. To rely on empirical analysis in the absence of theory is to rely solely on anecdotes, and from anecdotes it is possible to convince ourselves of almost anything.

Worse than an absence of theory, however, is incorrect theory that is accepted without sufficient scrutiny. Unfortunately, many of the early efforts to offer a theory that explains how evolutionary algorithms work were flawed; even more disappointing, the recognition that these efforts were not entirely sound is only recently coming to light. Only within the last five years has it been shown that (1) the k-armed bandit analysis offered in the mid-1970s was not well formed, (2) proportional selection does not generate an "optimal allocation of trials," (3) binary representations are not optimal for function optimization, (4) no specific cardinality of representation is generally superior, (5) evolutionary algorithms are not uniformly improved when recombination is employed, (6) the schema theorem may not apply when fitness is a random variable (and may be of little or no significance in explaining the trajectory of an evolutionary algorithm over multiple generations regardless), and (7) self-adaptation does not require recombination to be successful. Certainly this listing could be extended. Too many early speculations were accepted without sufficient scrutiny; more skepticism and caution need to be applied.

The no free lunch theorem applies to algorithms that do not resample points. In contrast, evolutionary algorithms have the potential to resample points. Obviating this possibility would appear to require searching a growing list of points generated at each iteration and eliminating steps that would revisit old points. The computational overhead of this approach is not attractive. Thus, it becomes important to ask about the likelihood of resampling. Salomon (1997) showed that, in the case of, say, continuous function optimization, when an evolutionary algorithm varies all parameters simultaneously via multivariate Gaussian perturbation, the factor of resampling is of small order; however, when each parameter is varied with only a small probability, as is often the case when a high reliance is placed on recombination and a low reliance is placed on mutation, this factor can grow as $O(n \log n)$. As a result, such designs can perform worse than a random search, on average, when it is constrained not to resample. That is, some standard implementations of evolutionary algorithms may be worse than enumeration across all problems.

In some cases, empirical trials can lead to a greater understanding and theoretical derivations. For example, there has been recent interest in using Cauchy distributions instead of Gaussian distributions for varying continuous parameters (Yao and Liu, 1996, 1998; Kappler, 1996; Fogel et al., 1997). The observed results on several test functions and applications showed that evolutionary algorithms that rely on Cauchy distributions may have less tendency to become trapped in local optima because of their relatively fatter tails. The encouraging experimental results have led to some theoretical investigations. Rudolph (1997b) showed that the local convergence rates are smaller with Cauchy variations than with Gaussian variations. Yao et al. (1997) showed that when a step size larger than σ (the standard deviation for a Gaussian variation) is required, the Cauchy distribution will have greater probability of yielding that step. When smaller steps are beneficial, Gaussian variation will have a corresponding advantage. These results, in turn, have promoted interest in operators that blend (convolve) Gaussian and Cauchy variations (Chellapilla, 1998b) or self-adapt the use of one or the other (e.g., Saravanan and Fogel, 1997; and others). This is a good example of theory and experiment working with synergistic effects.

Ultimately, since no single choice of variation, selection, population size, and so forth can be best in general, there is always a need for tools to assess and trade off different parameter choices in specific circumstances. Fitness distributions offer one such tool. By examining the fitness of offspring as the result of applying different mutations, recombinations, and so forth, it is possible to identify choices that lead to statistically significantly better results in specific circumstances. These fitness distributions can also provide feedback on whether or not an intuition concerning the utility of a specific variation or other operator is correct. Determining the appropriate design choices by classes of function, initialization, representation, and so forth, remains an open question—one that cannot be answered by examining any of these facets in isolation.

REFERENCES

Ackley, D. H. (1987). *A Connectionist Machine for Genetic Hill-Climbing*. Boston: Kluwer Academic Publishers.

Altenberg, L. (1995). "The Schema Theorem and Price's Theorem," *Foundations of Genetic Algorithms 3*, edited by L. D. Whitley and M. D. Vose, San Mateo, CA: Morgan Kaufmann, pp. 23–49.

Angeline, P. J. (1996). "The Effects of Noise on Self-Adaptive Evolutionary Optimization," In *Evolutionary Programming V: Proc. of the 5th Annual Conference on Evolutionary Programming*, edited by L. J. Fogel, P. J. Angeline, and T. Bäck, Cambridge, MA: MIT Press, pp. 433–439.

Angeline, P. J. (1997). "Subtree Crossover: Building Block Engine or Macromutation?" In *Genetic Programming 1997: Proc. of the Second Annual Conference*, edited by J. R. Koza, K. Deb, M. Dorigo, D. B. Fogel, M. Garzon, H. Iba, and R. L. Riolo, San Mateo, CA: Morgan Kaufmann, pp. 9–17.

Atmar, W. (1990). "Natural Processes Which Accelerate the Evolutionary Search," In *Proc. of 24th Asilomar Conf. on Signals, Systems and Computers,* edited by R. R. Chen, Pacific Grove, CA: Maple Press, pp. 1030–1035.

Bäck, T. (1992). "The Interaction of Mutation Rate, Selection, and Self-Adaptation Within a Genetic Algorithm," In *Parallel Problem Solving from Nature 2,* edited by R. Männer and B. Manderick, Amsterdam: North-Holland, pp. 85–94.

Bäck, T. (1993). Personal Communication, Univ. of Dortmund, Germany.

Bäck, T. (1994). "Selective Pressure in Evolutionary Algorithms: A Characterization of Selection Methods," In *Proc. of the First IEEE Conference on Evolutionary Computation,* Piscataway, NJ: IEEE Press, pp. 57–62.

Bäck, T. (1996). *Evolutionary Algorithms in Theory and Practice.* New York: Oxford.

Bäck, T., F. Hoffmeister, and H.-P. Schwefel (1991). "A Survey of Evolution Strategies," In *Proc. of the Fourth Intern. Conf. on Genetic Algorithms,* edited by R. K. Belew and L. B. Booker, San Mateo, CA: Morgan Kaufmann, pp. 2–9.

Bäck, T., G. Rudolph, and H.-P. Schwefel (1993). "Evolutionary Programming and Evolution Strategies: A Comparison," In *Proc. of 2nd Annual Conference on Evolutionary Programming,* edited by D. B. Fogel and W. Atmar, La Jolla, CA: Evolutionary Programming Society, pp. 11–22.

Banzhaf, W., P. Nordin, R. E. Keller, and F. D. Francone (1998). *Genetic Programming: An Introduction,* San Francisco: Morgan Kaufmann.

Beyer, H.-G. (1994). "Toward a Theory of Evolution Strategies: The (μ,λ)-Theory," *Evolutionary Computation,* Vol. 2:4, pp. 381–407.

Beyer, H.-G. (1995). "Toward a Theory of Evolution Strategies: On the Benefits of Sex—the $(\mu/\mu,\lambda)$-Theory," *Evolutionary Computation,* Vol. 3:1, pp. 81–111.

Beyer, H.-G., and D. B. Fogel (1997). "A Note on the Escape Probabilities for Two Alternative Methods of Selection under Gaussian Mutation," In *Evolutionary Programming VI,* edited by P. J. Angeline, R. G. Reynolds, J. R. McDonnell, and R. Eberhart (eds.), Berlin: Springer, pp. 265–274.

Bohachevsky, I. O., M. E. Johnson, and M. L. Stein (1986). "Generalized Simulating Annealing for Function Optimization," *Technometrics,* Vol. 28:3, pp. 209–218.

Bremermann, H. J. (1958). "The Evolution of Intelligence. The Nervous System as a Model of Its Environment," Technical Report No. 1, Contract No. 477(17), Dept. of Mathematics, University of Washington, Seattle, July.

Bremermann, H. J. (1962). "Optimization Through Evolution and Recombination," In *Self-Organizing Systems,* edited by M. C. Yovits, G. T. Jacobi, and G. D. Goldstine, Washington, DC: Spartan Books, pp. 93–106.

Chellapilla, K. (1997). "Evolving Computer Programs Without Subtree Crossover," *IEEE Trans. Evolutionary Computation,* Vol. 1:3, pp. 209–216.

Chellapilla, K. (1998a). "A Preliminary Investigation into Evolving Modular Programs without Subtree Crossover," In *Genetic Programming 98: Proc. of the Third Ann. Genetic Programming Conf.,* edited by J. R. Koza, W. Banzhaf, K. Chellapilla, K. Deb, M. Dorigo, D. B. Fogel, M. H. Garzon, D. E. Goldberg, H. Iba, and R. L. Riolo, San Francisco: Morgan Kaufmann, pp. 23–31.

Chellapilla, K. (1998b). "Combining Mutation Operators in Evolutionary Programming," *IEEE Trans. Evolutionary Computation,* Vol. 2:3, pp. 91–96.

Davis, L. (1985). "Applying Adaptive Algorithms to Epistatic Domains," In *Proc. of Intern. Joint Conf. on Art. Intel.,* chaired by A. Joshi, Los Altos, CA: Morgan Kaufmann, pp. 162–164.

Davis, L. (1989). "Adding Operator Probabilities in Genetic Algorithms," In *Proc. of the Third International Conference on Genetic Algorithms,* edited by J. D. Schaffer, San Mateo, CA: Morgan Kaufmann, pp. 61–69.

Davis, L. (ed.) (1991). *Handbook of Genetic Algorithms,* New York: Van Nostrand Reinhold.

Davis, T. E., and J. C. Principe (1993). "A Markov Chain Framework for the Simple Genetic Algorithm," *Evolutionary Computation,* Vol. 1:3, pp. 269–288.

De Jong, K. A. (1975). "The Analysis of the Behavior of a Class of Genetic Adaptive Systems," Doctoral Dissertation, Univ. of Michigan, Ann Arbor.

De Jong, K. A., W. M. Spears, and D. F. Gordon (1995). "Using Markov Chains to Analyze GAFOs," In *Foundations of Genetic Algorithms 3,* edited by L. D. Whitley and M. D. Vose, San Mateo, CA: Morgan Kaufmann, pp. 115–137.

Eiben, A. E., E. H. L. Aarts, K. M. Van Hee (1991). "Global Convergence of Genetic Algorithms: A Markov Chain Analysis," In *Parallel Problem Solving from Nature,* edited by H.-P. Schwefel and R. Männer, Berlin: Springer, pp. 4–12.

English, T. M. (1996). "Evaluation of Evolutionary and Genetic Optimizers: No Free Lunch," *In Evolutionary Programming V: Proc. of the 5th Annual Conference on Evolutionary Programming,* edited by L. J. Fogel, P. J. Angeline, and T. Bäck, Cambridge, MA: MIT Press, pp. 163–169.

Fang, K.-T., S. Kotz, and K.-W. Ng (1990). *Symmetric Multivariate and Related Distributions.* Vol. 36 of Monographs on Statistics and Applied Probability, London: Chapman and Hall.

Fisher, R. A. (1930). *The Genetical Theory of Natural Selection.* Oxford: Clarendon Press.

Fogel, D. B. (1992). "Evolving Artificial Intelligence," Doctoral Dissertation, UCSD.

Fogel, D. B. (1994a). "Asymptotic Convergence Properties of Genetic Algorithms and Evolutionary Programming: Analysis and Experiments," *Cybernetics and Systems,* Vol. 25:3, pp. 389–407.

Fogel, D. B. (1994b). "Applying Evolutionary Programming to Selected Control Problems," *Computers Math. Applic.,* Vol. 27:11, pp. 89–104.

Fogel, D. B. (1995). "Phenotypes, Genotypes, and Operators," In *Proc. of 1995 IEEE Conference on Evolutionary Computation,* Piscataway, NJ: IEEE Press, pp. 193–198.

Fogel, D. B., and J. W. Atmar (1990). "Comparing Genetic Operators with Gaussian Mutations in Simulated Evolutionary Processes Using Linear Systems," *Biological Cybernetics,* Vol. 63, pp. 111–114.

Fogel, D. B., L. J. Fogel, and J. W. Atmar (1991). "Meta-Evolutionary Programming,"

Proc. of 25th Asilomar Conf. on Signals, Systems and Computers, edited by R. R. Chen, Maple Press, Pacific Grove, CA, pp. 540–545.

Fogel, D. B., L. J. Fogel, W. Atmar, and G. B. Fogel (1992). "Hierarchic Methods of Evolutionary Programming," In *Proc. of the First Ann. Conf. on Evolutionary Programming,* edited by D. B. Fogel and W. Atmar, La Jolla, CA: Evolutionary Programming Society, pp. 175–182.

Fogel, D. B., and A. Ghozeil (1996). "Using Fitness Distributions to Design More Efficient Evolutionary Computations," In *Proc. of 1996 IEEE Conference on Evolutionary Computation,* Piscataway, NJ: IEEE Press, pp. 11–19.

Fogel, D. B., and A. Ghozeil (1997a). "Schema Processing under Proportional Selection in the Presence of Random Effects," *IEEE Trans. Evolutionary Computation,* Vol. 1:4, pp. 290–293.

Fogel, D. B., and A. Ghozeil (1997b). "A Note on Representations and Variation Operators," *IEEE Trans. Evolutionary Computation,* Vol. 1:2, pp. 159–161.

Fogel, D. B., and A. Ghozeil (1998). "The Schema Theorem and the Misallocation of Trials in the Presence of Stochastic Effects," In *Evolutionary Programming VII,* edited by V. W. Porto, N. Saravanan, D. Waagen, and A. E. Eiben, Berlin: Springer, pp. 313–321.

Fogel, D. B., and L. C. Stayton (1994). "On the Effectiveness of Crossover in Simulated Evolutionary Optimization," *BioSystems,* Vol. 32:3, pp. 171–182.

Fogel, D. B., E. C. Wasson, E. M. Boughton, and V. W. Porto (1997). "A Step Toward Computer-Assisted Mammography Using Evolutionary Programming and Neural Networks," *Cancer Letters,* Vol. 119, pp. 93–97.

Fogel, G. B., and D. B. Fogel (1995). "Continuous Evolutionary Programming," *Cybernetics and Systems,* Vol. 26, pp. 79–90.

Fogel, L. J. (1964). "On the Organization of Intellect," Doctoral Dissertation, UCLA.

Fraser, A. S. (1957a). "Simulation of Genetic Systems by Automatic Digital Computers. I. Introduction," *Australian J. of Biol. Sci.,* Vol. 10, pp. 484–491.

Friedberg, R. M. (1958). "A Learning Machine: Part I," *IBM J. of Research and Development,* Vol. 2, pp. 2–13.

Fuchs, M. (1998). "Crossover versus Mutation: An Empirical and Theoretical Case Study," In *Genetic Programming 98: Proc. of the Third Ann. Genetic Programming Conf.,* edited by J. R. Koza, W. Banzhaf, K. Chellapilla, K. Deb, M. Dorigo, D. B. Fogel, M. H. Garzon, D. E. Goldberg, H. Iba, and R. L. Riolo, San Francisco: Morgan Kaufmann, pp. 78–85.

Goldberg, D. E. (1987). "Simple Genetic Algorithms and the Minimal, Deceptive Problem," In *Genetic Algorithms and Simulated Annealing,* edited by L. Davis, San Mateo, CA: Morgan Kaufmann, pp. 74–88.

Goldberg, D. E. (1989). *Genetic Algorithms in Search, Optimization and Machine Learning.* Reading, MA: Addison-Wesley.

Goldberg, D. E., and K. Deb (1991). "A Comparative Analysis of Selection Schemes Used in Genetic Algorithms," In *Foundations of Genetic Algorithms,* edited by G. J. E. Rawlins, San Mateo, CA: Morgan Kaufmann, pp. 69–93.

Goodman, R. (1988). *Introduction to Stochastic Models.* Reading, MA: Benjamin/ Cummings Publishing Co.

Grefenstette, J. J. (1986). "Optimization of Control Parameters for Genetic Algorithms," *IEEE Trans. on Systems, Man and Cybernetics,* Vol. SMC-16, No. 1, pp. 122–128.

Grefenstette, J. J. (1995). "Predictive Models Using Fitness Distributions of Genetic Operators," In *Foundations of Genetic Algorithms 3,* edited by L. D. Whitley and M. D. Vose, San Mateo, CA: Morgan Kaufmann, pp. 139–161.

Holland, J. H. (1973). "Genetic Algorithms and the Optimal Allocation of Trials," *SIAM J. Computing,* Vol. 2:2, pp. 88–105.

Holland, J. H. (1975). *Adaptation in Natural and Artificial Systems.* Ann Arbor, MI: Univ. of Michigan Press.

Holland, J. H. (1992). "Genetic Algorithms," *Scientific American,* July, pp. 66–72.

Hsu, P. L., and H. Robbins (1947). "Complete Convergence and the Law of Large Numbers," *Proc. Nat. Acad. Sci.,* Vol. 33:2, pp. 25–31.

Johnson, N. L., and S. Kotz (1970). *Distributions in Statistics: Continuous Distributions—2.* Boston: Houghton Mifflin.

Jones, T. (1995). "Crossover, Macromutation, and Population-based Search," In *Proc. of the 6th Intern. Conf. on Genetic Algorithms,* edited by L. Eshelman, San Mateo, CA: Morgan Kaufmann, pp. 73–80.

Kallel, L., and M. Schoenauer (1997). "A Priori Comparison of Binary Crossover Operators: No Universal Statistical Measure, But a Set of Hints," In *Artificial Evolution 1997,* edited by J.-K. Hao, E. Lutton, E. Ronald, M. Schoenauer, and D. Snyers, Berlin: Springer, pp. 287–299.

Kappler, C. (1996). "Are Evolutionary Algorithms Improved by Large Mutations?" In *Parallel Problem Solving from Nature 4,* edited by H.-M. Voigt, W. Ebeling, I. Rechenberg, and H.-P. Schwefel, Berlin: Springer, pp. 346–355.

Koza, J. R. (1992). *Genetic Programming.* Cambridge, MA: MIT Press.

Luke, S., and L. Spector (1998). "A Revised Comparison of Crossover and Mutation in Genetic Programming," In *Genetic Programming 98: Proc. of the Third Ann. Genetic Programming Conf.,* edited by J. R. Koza, W. Banzhaf, K. Chellapilla, K. Deb, M. Dorigo, D. B. Fogel, M.H. Garzon, D. E. Goldberg, H. Iba, and R. L. Riolo, San Francisco: Morgan Kaufmann, pp. 208–213.

Macready, W. G., and D. H. Wolpert (1998). "Bandit Problems and the Exploration/Exploitation Tradeoff," *IEEE Trans. Evolutionary Computation,* Vol. 2:1, pp. 2–22.

Nordin, P., and W. Banzhaf (1995). "Complexity Compression and Evolution," In *Proc. 6th Intern. Conf. on Genetic Algorithms,* edited by L. Eshelman, San Mateo, CA: Morgan Kaufmann, pp. 310–317.

Nordin, P., F. Francone, and W. Banzhaf (1996). "Explicitly Defined Introns and Destructive Crossover in Genetic Programming," In *Advances in Genetic Programming 2,* edited by P. J. Angeline and K. E. Kinnear, Cambridge, MA: MIT Press, pp. 111–134.

Oliver, I. M., D. J. Smith, and J.R.C. Holland (1987). "A Study of Permutation Crossover Operators on the Traveling Salesman Problem," In *Proc. 2nd Intern. Conf. on Genetic Algorithms,* edited by J. J. Grefenstette, San Mateo, CA: Morgan Kaufmann, pp. 224–230.

Papoulis, A. (1984). *Probability, Random Variables, and Stochastic Processes,* 2nd ed., New York: McGraw-Hill.

Rappl, G. (1984). "Konvergenzraten von Random Search Verfahren zur globalen Optimierung," Dissertation, HSBw München, Germany.

Rechenberg, I. (1965). "Cybernetic Solution Path of an Experimental Problem," Royal Aircraft Establishment, Library Translation No. 1122, August.

Rechenberg, I. (1973). *Evolutionsstrategie: Optimierung technischer Systeme nach Prinzipien der biologischen Evolution.* Stuttgart: Frommann-Holzboog Verlag.

Rechenberg, I. (1994). Personal Communication, Tech. Univ. Berlin, Germany.

Reed, J., R. Toombs, and N. A. Barricelli (1967). "Simulation of Biological Evolution and Machine Learning," *J. Theoretical Biology,* Vol. 17, pp. 319–342.

Rosenberg, R. (1967). "Simulation of Genetic Populations with Biochemical Properties, Doctoral Dissertation, Univ. Michigan, Ann Arbor.

Rudolph, G. (1994a). "Convergence Analysis of Canonical Genetic Algorithms," *IEEE Trans. on Neural Networks,* Vol. 5:1, pp. 96–101.

Rudolph, G. (1994b). "Massively Parallel Simulated Annealing and Its Relation to Evolutionary Algorithms," *Evolutionary Computation,* Vol. 1:4, pp. 361–383.

Rudolph, G. (1996). "Convergence Properties of Evolutionary Algorithms," Doctoral Dissertation, Univ. Dortmund, Germany.

Rudolph, G. (1997a). "Reflections on Bandit Problems and Selection Methods in Uncertain Environments," In *Proc. of 7th Intern. Conf. on Genetic Algorithms,* edited by T. Bäck, San Francisco: Morgan Kaufmann, pp. 166–173.

Rudolph, G. (1997b). "Local Convergence Rates of Simple Evolutionary Algorithms with Cauchy Mutations," *IEEE Trans. Evolutionary Computation,* Vol. 1:4, pp. 249–258.

Salomon, R. (1996). "Re-Evaluating Genetic Algorithm Performance under Coordinate Rotation of Benchmark Functions: A Survey of Some Theoretical and Practical Aspects of Genetic Algorithms," *BioSystems,* Vol. 39:3, pp. 263–278.

Salomon, R. (1997). "Raising Theoretical Questions about the Utility of Genetic Algorithms," In *Evolutionary Programming VI,* edited by P. J. Angeline, R. G. Reynolds, J. R. McDonnell, and R. Eberhart, Berlin: Springer, pp. 275–284.

Saravanan, N., and D. B. Fogel (1994). "Learning of Strategy Parameters in Evolutionary Programming: An Empirical Study," In *Proc. of the Third Ann. Conf. on Evolutionary Programming,* edited by A. V. Sebald and L. J. Fogel, River Edge, NJ: World Scientific, pp. 269–280.

Saravanan, N., and D. B. Fogel (1997). "Multi-operator Evolutionary Programming: A Preliminary Study on Function Optimization," In *Evolutionary Programming VI,* edited by P. J. Angeline, R. G. Reynolds, J. R. McDonnell, and R. Eberhart, Berlin: Springer, pp. 215–221.

Schaffer, J. D., R. A. Caruana, L. J. Eshelman, and R. Das (1989). "A Study of Control Parameters Affecting Online Performance of Genetic Algorithms for Function Optimization," In *Proc. of the Third Intern. Conf. on Genetic Algorithms,* edited by J. D. Schaffer (ed.), San Mateo, CA: Morgan Kaufmann, pp. 51–59.

Schaffer, J. D., and L. J. Eshelman (1991). "On Crossover as an Evolutionarily Viable Strategy," In *Proc. of the Fourth Intern. Conf. on Genetic Algorithms,* edited by R. K. Belew and L. B. Booker, San Mateo, CA: Morgan Kaufmann, pp. 61–68.

Schaffer, J. D., and A. Morishima (1987). "An Adaptive Crossover Distribution Mechanism for Genetic Algorithms," In *Proc. of the Second Intern. Conf. on Genetic Algorithms,* edited by J. J. Grefenstette, Hillsdale, NJ: Lawrence Erlbaum, pp. 36–40.

Scheel, A. (1985). "Beitrag zur Theorie der Evolutionssstrategie," Dissertation, Tech. Univ. Berlin, Berlin.

Schraudolph, N. N. (1992). Personal Communication, Salk Institute, La Jolla, CA.

Schraudolph, N. N., and R. K. Belew (1992). "Dynamic Parameter Encoding for Genetic Algorithms," *Machine Learning,* Vol. 9, pp. 9–21.

Schwefel, H.-P. (1977). *Numerische Optimierung von Computer-Modellen mittels der Evolutionsstrategie.* Interdisciplinary Systems Research, Vol. 26, Birkhäuser, Basel.

Schwefel, H.-P. (1981). *Numerical Optimization of Computer Models.* Chichester, U.K.: John Wiley.

Sebald, A. V. (1991). Personal Communication, UCSD.

Serfling, R. J. (1980). *Approximation Theorems of Mathematical Statistics.* New York: John Wiley.

Spears, W. M., and K. A. De Jong (1991a). "On the Virtues of Parameterized Uniform Crossover," In *Proc. of the 4th Intern. Conf. on Genetic Algorithms,* edited by R. K. Belew and L. B. Booker, San Mateo, CA: Morgan Kaufmann, pp. 230–236.

Spears, W. M., and K. A. De Jong (1991b). "An Analysis of Multi-Point Crossover," In *Foundations of Genetic Algorithms,* edited by G.J.E. Rawlins, San Mateo, CA: Morgan Kaufmann, pp. 301–315.

Spears, W. M., and K. A. De Jong (1997). "Analyzing GAs Using Markov Models with Semantically Ordered and Lumped States," In *Foundations of Genetic Algorithms 4,* edited by M. D. Vose and R. K. Belew, San Mateo, CA: Morgan Kaufmann, pp. 84–100.

Syswerda, G. (1989). "Uniform Crossover in Genetic Algorithms," In *Proc. of the Third Intern. Conf. on Genetic Algorithms,* edited by J. D. Schaffer, San Mateo, CA: Morgan Kaufmann, pp. 2–9.

Teller, A. (1996). "Evolving Programmers: The Coevolution of Intelligent Recombination Operators," In *Advances in Genetic Programming 2,* edited by P. J. Angeline and K. E. Kinnear, Cambridge, MA: MIT Press, pp. 45–68.

Whitley, L. D. (1993). "Introduction," In *Foundations of Genetic Algorithms 2,* edited by L. D. Whitley, San Mateo, CA: Morgan Kaufmann, pp. 1–4.

Wolpert, D. H., and W. G. Macready (1997). "No Free Lunch Theorems for Optimization," *IEEE Trans. Evolutionary Computation,* Vol. 1:1, pp. 67–82.

Yao, X., L. Guangming, and Y. Liu (1997). "An Analysis of Evolutionary Algorithms Based on Neighborhood and Step Sizes," In *Evolutionary Programming VI*, edited by P. J. Angeline, R. G. Reynolds, J. R. McDonnell, and R. Eberhart, Berlin: Springer, pp. 297–307.

Yao, X., and Y. Liu (1996). "Fast Evolutionary Programming," In *Evolutionary Programming V*, edited by L. J. Fogel, P. J. Angeline, and T. Bäck, Cambridge, MA: MIT Press, pp. 451–460.

Yao, X., and Y. Liu (1998). "Scaling Up Evolutionary Programming Algorithms," In *Evolutionary Programming VII*, edited by V. W. Porto, N. Saravanan, D. Waagen, and A. E. Eiben, Berlin: Springer, pp. 103–112.

CHAPTER 5

INTELLIGENT BEHAVIOR

5.1 INTELLIGENCE REQUIRES ITERATIVE PREDICTION AND CONTROL

The process of adaptation is one of minimizing surprise to the adaptive organism. It requires three basic elements: prediction, control, and feedback. Prediction operates as a mechanistic mapping from a set of observed environmental symbols to another set of symbols that represents the expected new circumstance. The mapping is essentially a model that relates previous experiences to future outcomes. Prediction is an essential ingredient of intelligence, for if a system cannot predict future events, every environmental occurrence comes as a surprise and adaptation is impossible. But an intelligent system must not simply predict its environment; rather, it must use its predictions to affect its decision making and thereby control its behavior. Any system that allocates its resources (i.e., controls its behavior) without regard to the anticipated consequences of those actions relegates its future to nothing but pure luck. Furthermore, the adaptive organism must act on the error in prediction and the associated cost of inappropriate behaviors to improve the quality of its forecasting. Every intelligent system adapts to its environment by predicting future events, controlling its actions in light of those predictions, and revising its bases for making predictions in light of feedback on the degree to which it is achieving its goals.

The examples offered in this chapter illustrate this fundamental concept. The chosen examples highlight the ability of evolutionary computation to (1) induce predictive models of systems and to use such models for control in light of a purpose, (2) adapt behavior while interacting with other purposive agents, and (3) achieve success in problem solving without relying on credit assignment algorithms. These cases are not offered as exhaustive evidence of the range of possible demonstrations, but instead are meant to serve as a foundation for the thesis that machine intelligence can be achieved by simulating evolution, and that such intelligence can be used for good engineering purpose. In each case, the evolutionary simulation iteratively improves algorithms that operate on experienced data and generate increasingly appropriate behaviors in the light of environmental demands.

5.2 GENERAL PROBLEM SOLVING: EXPERIMENTS WITH TIC-TAC-TOE

To provide a simple example of evolutionary problem solving, an evolutionary program was written to discover optimal strategies in tic-tac-toe (Fogel, 1992, 1993b). The game is well known, but it will be described in detail for completeness. There are two players and a three-by-three grid (Figure 5-1). Initially, the grid is empty. Each player moves in turn by placing a marker in an open square. By convention, the first player's marker is "X" and the second player's marker is "O." The first player moves first. The object of the game is to place three markers in a row. This results in a win for that player and a loss for the opponent. Failing a win, a draw may be earned by preventing the opponent from placing three markers in a row. It can be shown by enumerating the game tree that at least a draw can be forced by the second player.

Attention was devoted to evolving a strategy for the first player (an equivalent procedure could be used for the second player). A suitable coding structure had to be selected. It had to receive a board pattern as input and yield a corresponding move as output. The coding structure utilized in these experiments was a multilayered feed forward perceptron (Figure 5-2). Each hidden or output node performed a sum of the weighted input strengths, subtracted off an adaptable bias term, and passed the result through a sigmoid filter, $(1 + e^{-x})^{-1}$. Only a single hidden layer was incorporated. This architecture was selected because (1) variations are universal function approximators, (2) the response to any stimulus could be evaluated rapidly, and (3) the extension to multiple hidden layers is obvious.

There were nine input and output units. Each corresponded to a square in the grid. An "X" was denoted by the value 1.0, an "O" was denoted by the value -1.0, and an open space was denoted by the value 0.0. A move was determined by presenting the current board pattern to the network and examining the relative strengths of the nine output nodes. A marker was placed in the empty square with the maximum output strength. This procedure guaranteed

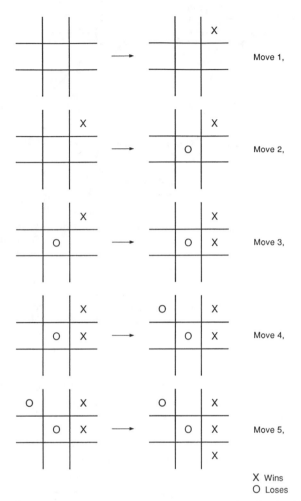

Move 1,

Move 2,

Move 3,

Move 4,

Move 5,

X Wins
O Loses

Figure 5-1 A possible game of tic-tac-toe. Players alternate moves in a 3 × 3 grid. The first player to place three markers in a row wins.

legal moves. The output from nodes associated with squares in which a marker had already been placed was ignored. No selection pressure was applied to drive the output from such nodes to zero.

The initial population consisted of 50 parent networks. The number of nodes in the hidden layer was chosen at random in accordance with a uniform distribution over the integers [1, . . . , 10]. The initial weighted connection

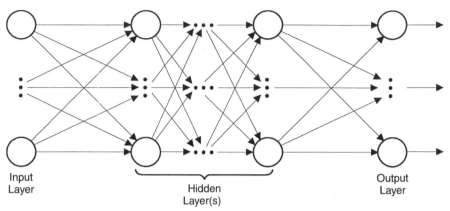

Figure 5-2 A feed forward neural network. A number of input nodes pass information from the environment to nodes in a potential series of hidden layers. The final hidden layer is connected to the output layer, which yields a response of the network to any input pattern. In a multilayer perceptron, each node typically performs a dot product of the inputs to the node and their respective weights, subtracts off a bias term, and then passes the result through a nonlinear filter of the form $[1 + \exp(-x)]^{-1}$, where x is the result before filtering. The networks in the current experiments used a feed forward neural network with nine input and output nodes and only a single hidden layer of a variable number of nodes.

strengths and bias terms were randomly distributed according to a uniform distribution ranging over $[-0.5, 0.5]$. A single offspring was copied from each parent and modified by two modes of mutation:

1. All weight and bias terms were perturbed by adding a Gaussian random variable with zero mean and a standard deviation of 0.05.
2. With a probability of 0.5, the number of nodes in the hidden layer was allowed to vary. If a change was indicated, there was an equal likelihood that a node would be added or deleted, subject to the constraints on the maximum and minimum number of nodes (10 and one, respectively). Nodes to be added were initialized with all weights and the bias term being set equal to 0.0.

A rule-based procedure that played nearly perfect tic-tac-toe was implemented to evaluate each contending network (see below). The execution time with this format was linear with the population size and provided the opportunity for multiple trials and statistical results. The evolving networks were allowed to move first in all games. The first move was examined by the rule base with the eight possible second moves being stored in an array. The rule base proceeded as follows:

1. From the array of all possible moves, select a move that has not yet been played.
2. For subsequent moves:
   (a) with a 10 percent chance, move randomly, else
   (b) if a win is available, place a marker in the winning square, else
   (c) if a block is available, place a marker in the blocking square, else
   (d) if two open squares are in line with an "O," randomly place a marker in either of the two squares, else
   (e) randomly move in any open square.
3. Continue with step 2 until the game is completed.
4. Continue with step 1 until games with all eight possible second moves have been played.

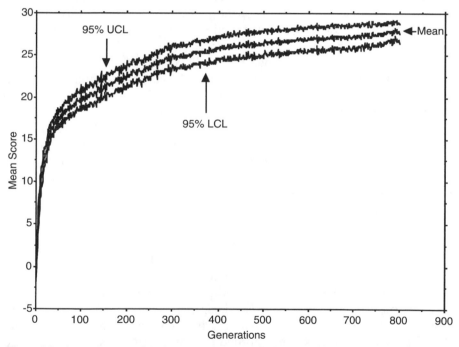

Figure 5-3 The mean score of the best neural network in the population as a function of the number of generations averaged across all 30 trials using payoffs of $+1$, -1, and 0 for winning, losing, and playing to a draw, respectively. The 95 percent upper and lower confidence limits (UCL and LCL) on the mean are calculated using a t-distribution.

The 10 percent chance for moving randomly was incorporated to maintain variety of play in an analogous manner to a persistence of excitation condition (Anderson et al., 1986, p. 40; Ljung, 1987, pp. 362–363). This feature and the restriction that the rule base only looks one move ahead makes the rule base nearly perfect, but beatable.

Each network was evaluated over four sets of these eight games. The payoff function varied in three sets of experiments over $\{+1, -1, 0\}$, $\{+1, -10, 0\}$ and $\{+10, -1, 0\}$, where the entries are the payoff for winning, losing, and playing to a draw, respectively. The maximum possible score over any four sets of games was 32 under the first two payoff functions and 320 under the latter payoff function. But a perfect score in any generation did not necessarily indicate a perfect algorithm because of the random variation in play generated by step 2a, above. After competition against the rule base was completed for all networks in the population, a second competition was held in which each network was compared with 10 other randomly chosen networks. If the score of the chosen network was greater than or equal to its competitor, it received a win. Those networks with the greatest number of wins were retained to be parents of the successive generations. Thirty trials were conducted with each payoff function. Evolution was halted after 800 generations in each trial.

The learning rate when using $\{+1, -1, 0\}$ was generally consistent across all trials (Figure 5-3). The 95 percent confidence limits around the mean were close to the average performance. Figure 5-4 depicts the improvement of the best player in the population at each generation in trial 2. The final best network in this trial possessed nine hidden nodes. The highest score achieved in trial 2 was 31. Figure 5-5 indicates the tree of possible games if the best-evolved network were played against the rule-based algorithm, omitting the possibility for random moves in step 2a. Under such conditions, the network would force a win in four of the eight possible branches of the tree and has the possibility of losing in two of the branches.

The learning rate when using $\{+1, -10, 0\}$ was considerably different (Figure 5-6) from that indicated in Figure 5-3. Selection in light of the increased penalty for losing resulted in an increased initial rate of improvement. Strategies that lose were quickly purged from the evolving population. After this first-order condition was satisfied, optimization continued to sort out strategies with the greatest potential for winning rather than drawing. Again, the 95 percent confidence limits around the mean were close to the average performance across all 30 trials. Figure 5-7 indicates the improvement of the best player in the population in trial 1. This network possessed 10 hidden

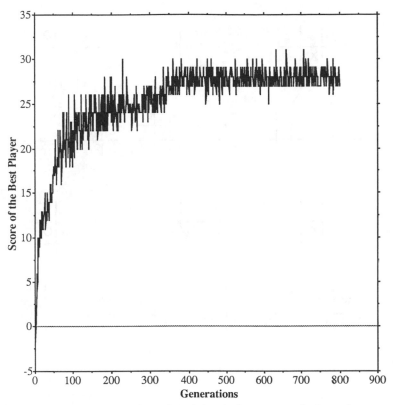

Figure 5-4 A typical learning curve (trial 2) depicting the improvement in the performance of the best player in the population as a function of the number of generations when the payoffs were +1, −1, and 0 for a win, loss, or draw, respectively.

nodes. Figure 5-8 depicts the tree of possible games if the best network from trial 1 were played against the rule-based player, omitting random moves from step 2a. It would not force a win in any branch of the tree. But it would also never lose. The payoff function was clearly reflected in the final observed behavior of the evolved network.

The learning rate when using {+10, −1, 0} appeared similar to that obtained when using {+1, −1, 0} (Figure 5-9). The 95 percent confidence limits were wider than was observed in the previous two experiments. The variability of the score was greater because a win received 10 more points than a draw, rather than only a single point more. Strategies with a propensity to lose were purged early in the evolution. By the 800th generation, most surviving strate-

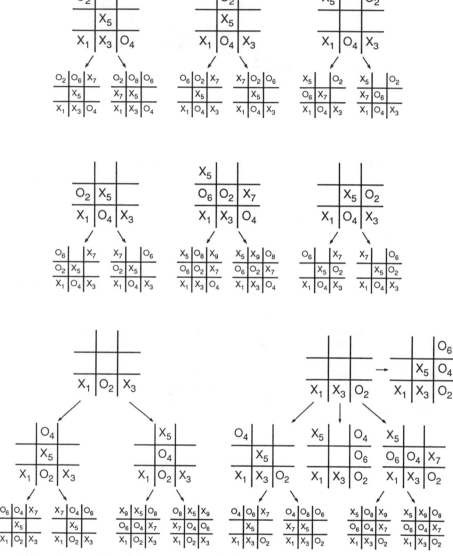

Figure 5-5 The tree of possible games when playing the best-evolved network from trial 2 with payoffs of +1, −1, and 0, for winning, losing, and playing to a draw, against the rule base without the 10 percent chance for making purely random moves. Each of the eight main boards corresponds to the eight possible next moves after the neural network moves first (in the lower left-hand corner). Branches indicate more than one possible move for the rule base, in which case all possibilities are depicted. The subscripts indicate the order of play.

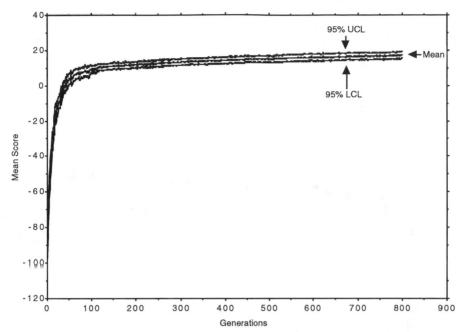

Figure 5-6 The mean score of the best neural network in the population as a function of the number of generations averaged across all 30 trials using payoffs of $+1$, -10, and 0 for winning, losing, and playing to a draw, respectively. The 95 percent upper and lower confidence limits (UCL and LCL) on the mean are calculated using a t-distribution. The
rate of optimization initially appears more rapid than is achieved when the penalty for
losing is merely equal to the gain for winning.

gies lost infrequently and varied mostly by their ability to win or draw. Figure
5-10 indicates the improvement of the best score in the population in trial 1.
The final best network possessed nine hidden nodes. The maximum possible
score is 320 (a win in each game). This was achieved once, just before the 800th
generation. The tree of possible games against the rule-based player is shown
in Figure 5-11. The best-evolved network in trial 1 could not achieve a perfect
score except when the rule-based procedure made errors in play through random perturbation (step 2a).

These results indicate a capability for problem solving. Behavior was
adapted to meet varying goals across a range of possible environments as determined by random chance moves on the part of the rule base. No a priori
information regarding the object of the game was offered to the evolutionary program. No hints regarding appropriate moves were given, nor were
there any attempts to assign values to various board patterns. The final out-

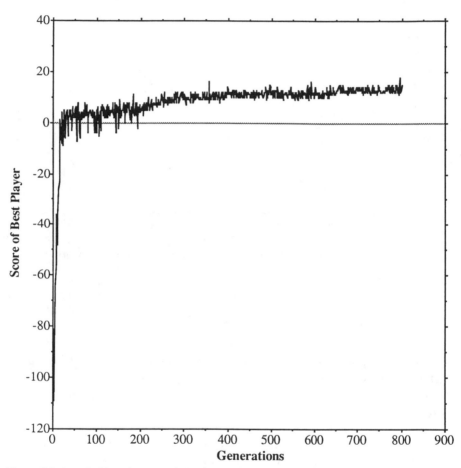

Figure 5-7 A typical learning curve (trial 1) depicting the improvement in the performance of the
best player in the population as a function of the number of generations when the payoffs were +1, −10, and 0 for a win, loss, or draw, respectively. The initial learning curve
appears steeper than that observed in Figure 5-4.

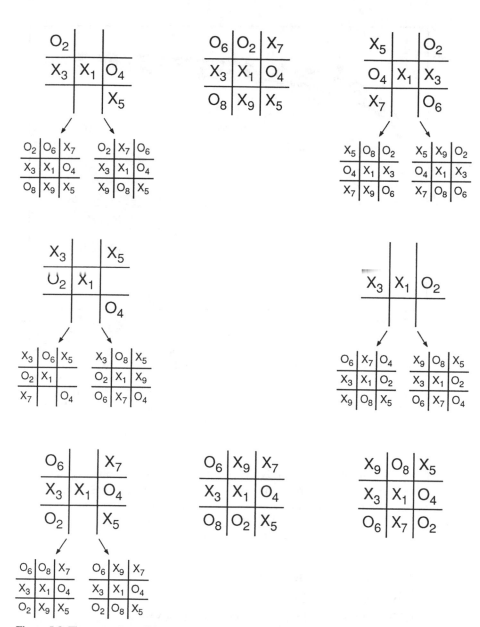

Figure 5-8 The tree of possible games when playing the best-evolved network from trial 1 with payoffs of +1, −10, and 0, for winning, losing, and playing to a draw, against the rule base without the 10 percent chance for making purely random moves. The format follows Figure 5-5. The best-evolved network never forces a win, but also never loses. This reflects the increased penalty for losing.

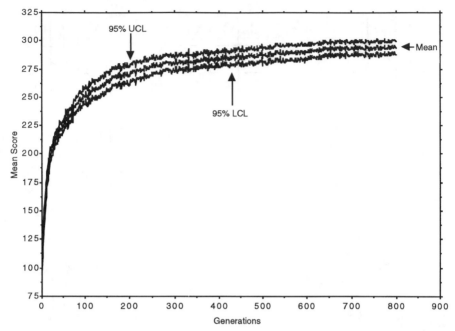

Figure 5-9 The mean score of the best neural network in the population as a function of the number of generations averaged across all 30 trials using payoffs of $+10$, -1, and 0 for winning, losing, and playing to a draw, respectively. The 95 percent upper and lower confidence limits (UCL and LCL) on the mean are calculated using a t-distribution. The
rate of optimization initially appears more similar to that which is achieved when the
penalty for losing is merely equal to the gain for winning.

come (win, lose, draw) was the only available information regarding the quality of play. Further, this information was only provided after 32 games had
been played; the contribution to the overall score from any single game was
not discernible. Heuristics regarding the environment were limited to the following:

1. There were nine inputs.
2. There were nine outputs.
3. Markers could only be placed in empty squares.

Essentially then, the procedure was only given an appropriate set of sen-
sors and a means for generating all possible behaviors in each setting and was

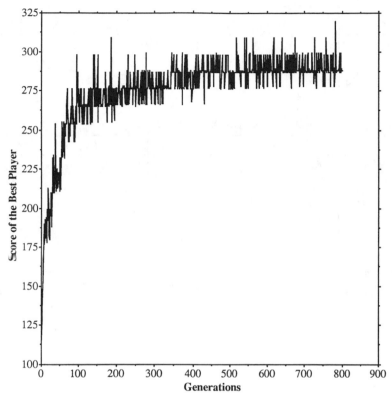

Figure 5-10 A typical learning curve (trial 1) depicting the improvement in the performance of the best player in the population as a function of the number of generations when the payoffs were +10, −1, and 0 for a win, loss, or draw, respectively. The learning curve appears similar to that observed in Figure 5-4, with an increased variance reflecting the larger gain for winning.

restricted to act within the physics of the environment (i.e., the rules of the game).

Evolutionary computation was able to adjust both the architecture and connections of the single hidden-layer perceptron simultaneously. The final evolved neural networks operated on each in a series of observed board patterns and in turn generated a sequence of generally appropriate responses (behaviors) in light of the specified goal. The final expressed behavior was clearly dependent on the payoff function. This evolutionary procedure is robust, and no fundamental alteration of the basic algorithm is required to address problems of arbitrary complexity.

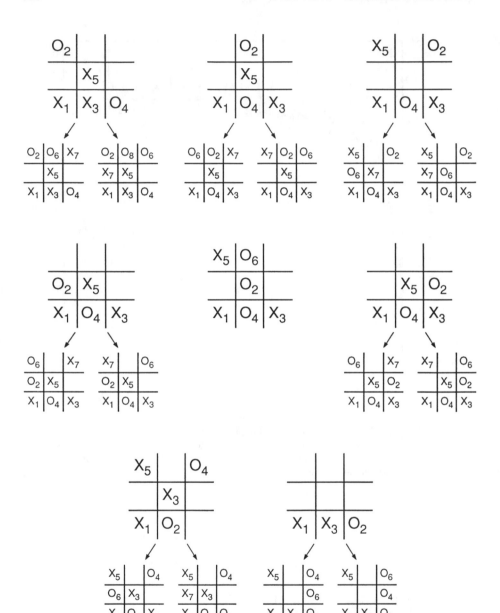

Figure 5-11 The tree of possible games when playing the best-evolved network from trial 1 with payoffs of +10, −1, and 0, for winning, losing, and playing to a draw, against the rule base without the 10 percent chance for making purely random moves. The format follows Figure 5-5. The game tree has many of the same characteristics observed in Figure 5-5, indicating little difference in resulting behavior when increasing the reward for winning as opposed to increasing the penalty for losing.

5.3 THE PRISONER'S DILEMMA: COEVOLUTIONARY ADAPTATION

The conditions that foster the evolution of cooperative behaviors among individuals of a species (or among species) are not generally well understood. In intellectually advanced social animals, cooperation between individuals, when it exists, is often ephemeral and quickly reverts to selfishness, with little or no clear indication of the specific circumstances that prompt the change. Simulation games such as the prisoner's dilemma have been used to gain insight into the precise conditions that promote the evolution of either decisively cooperative or selfish behavior in a community of individuals. Even simple games often generate very complex and dynamic optimization surfaces. The computational problem is to determine reliably any ultimately stable strategy (or strategies) for a specific game situation.

The prisoner's dilemma is an easily defined nonzero-sum, noncooperative game. The term *nonzero-sum* indicates that whatever benefits accrue to one player do not necessarily imply similar penalties imposed on the other player. The term *noncooperative* indicates that no preplay communication is permitted between the players. Typically, two players are involved in the game, each having two alternative actions: cooperate or defect. Cooperation implies increasing the total gain of both players; defecting implies increasing one's own reward at the expense of the other player. The optimal policy for a player depends on the policy of the opponent (Hofstadter, 1985, p. 717). Against a player who always defects, defection is the only rational play. But it is also the only rational play against a player who always cooperates, for such a player is a fool. Only when there is some mutual trust between the players does cooperation become a reasonable policy.

The general form of the game is represented in Table 5-1 (after Scodel et al., 1959). The game is conducted on a trial-by-trial basis (a series of moves). Each player must choose to cooperate or defect on each trial. The payoff matrix defining the game is subject to the following constraints (Rapoport, 1966):

$$2\gamma_1 > \gamma_2 + \gamma_3 \tag{5-1}$$

$$\gamma_3 > \gamma_1 > \gamma_4 > \gamma_2. \tag{5-2}$$

The first rule prevents any obvious desire for mutual cooperation. The second rule motivates each player to play noncooperatively.

Defection dominates cooperation in a game theoretic sense (Rapoport, 1966). But joint defection results in a payoff, γ_4, to each player that is smaller than the payoff, γ_1, that could be gained through mutual cooperation. If the game is played for only a single trial, the clearly compelling behavior is defec-

Table 5-1 The General Form of The Payoff Function in the
Prisoner's Dilemma, Where γ_1 is the Payoff to Each Player for
Mutual Cooperation, γ_2 is the Payoff for Cooperating When the
Other Player Defects, γ_3 is the Payoff for Defecting When the Other
Player Cooperates, and γ_4 is the Payoff for Mutual Defection. An
Entry (α, β) Indicates the Payoffs to Players A and B, Respectively.

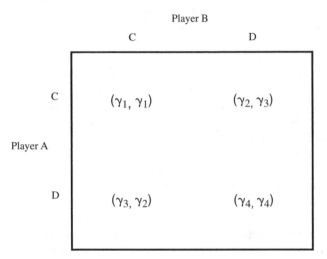

tion. If the game is iterated over many trials, the correct behavior becomes less
clear. If there is no possibility for communication with the other player, then
the iterated game degenerates into a sequence of independent trials, with defection again yielding the maximum expected benefit. But if the outcome of
earlier trials can affect the decision-making policy of the players in subsequent
trials, learning can take place and both players may seek to cooperate.

5.3.1   Background

In 1979, Axelrod organized a prisoner's dilemma tournament and solicited
strategies from game theorists who had published in the field (Axelrod,
1980a). The 14 entries were competed, along with a 15th entry: on each move,
cooperate or defect with equal probability. Each strategy was played against
all others and itself over a sequence of 200 moves. The specific payoff function used is shown in Table 5-2. The winner of the tournament, submitted by
Anatol Rapoport, was "Tit-for-Tat":

1. Cooperate on the first move.
2. Otherwise, mimic whatever the other player did on the previous move.

Table 5-2 The Specific Payoff Function Used in Axelrod (1980a).

Player B

		C	D
	C	(3, 3)	(0, 5)
Player A			
	D	(5, 0)	(1, 1)

Subsequent analysis by Axelrod (1984), Hofstadter (1985, pp. 721–723), and others indicates that this Tit-for-Tat strategy is robust because it never defects first and is never taken advantage of for more than one iteration at a time. Boyd and Lorberbaum (1987) indicated that Tit-for-Tat is not evolutionarily stable (*sensu* Maynard Smith, 1982). Nevertheless, in a second tournament, Axelrod (1980b) collected 62 contending entries, and again the winner was Tit-for-Tat.

Axelrod (1984) noted that eight of the 62 entries in the second tournament can be used to reasonably account for how well a given strategy did with the entire set. Axelrod (1987) used these eight strategies as opponents for a simulated evolving population of policies based on a genetic algorithm approach. Consideration was given to the set of strategies that is deterministic and uses outcomes of the three previous moves to determine a current move. Because there were four possible outcomes for each move, there were 4^3 or 64 possible sets of three possible moves. The coding for a policy was therefore determined by a string of 64 bits, where each bit corresponded with a possible instance of the preceding three interactions, and six additional bits that defined the player's move for the initial combinations of under three interactions. Thus there were 2^{70} (about 10^{21}) possible strategies.

The simulation was conducted as a series of steps: (1) Randomly select an initial population of 20 strategies; (2) execute each strategy against the eight representatives and record a weighted-average payoff; (3) determine the number of offspring from each parent strategy in proportion to their effectiveness;

(4) generate offspring by crossing over two parent strategies and, with a small probability, affect a mutation by randomly changing components of the strategy; and (5) continue to iterate this process. Crossover and mutation probabilities averaged one crossover and one-half a mutation per generation. Each game consisted of 151 moves (the average of the previous tournaments). A run consisted of 50 generations. Forty trials were conducted. From a random start, the technique created populations whose median performance was just as successful as Tit-for-Tat. The behavior of many of the strategies actually resembled Tit-for-Tat (Axelrod, 1987).

Another experiment (Axelrod, 1987) required the evolving policies to play against each other, rather than against the eight representatives. This was a much more complex environment: The opponents that each individual faced were concurrently evolving. As more effective strategies propagated throughout the population, each individual had to keep pace or face elimination through selection. Ten trials were conducted with this format. Typically, the population evolved away from cooperation initially, but then tended toward reciprocating whatever cooperation could be found. The average score of the population increased over time as "an evolved ability to discriminate between those who will reciprocate cooperation and those who won't" was attained (Axelrod, 1987).

5.3.2 Evolving Finite-State Representations

Fogel (1993a), following earlier investigation (Fogel, 1991), implemented a population of 100 coevolving finite-state machines (FSMs) (50 parents), each possessing a maximum of eight states. This limit was chosen to provide a reasonable chance for explaining the behavior of any machine through examination of its structure. Eight-state FSMs do not subsume all the behaviors that could be generated under the coding of Axelrod (1987), but do allow for a dependence on sequences of greater than third-order. The input alphabet consisted of {(C, C), (C, D), (D, C), (D, D)}, where the first symbol indicated the FSM's previous move and the second symbol indicated the opponent's previous move. The output alphabet consisted of {C, D}. Each FSM represented a predictive algorithm that operated on a sequence of stimulus pairs and generated a behavior in light of the expected next response of the opponent and the payoff for alternative interactions. Mutation was conducted as in Fogel et al. (1966)—that is, change an output symbol, state-transition, or start state, add a state, or delete a state—with an additional mutation operation being defined on the start symbol associated with each machine (i.e., will the machine cooperate or defect on the first move?). The initial machines possessed one to five states, with all responses and connections randomly determined. Each member of the population faced every other member in competition (round-robin). Scoring was made with respect to the payoff function in Axelrod (1987); refer to Table 5-2. Each game consisted of 151 moves, with the maximum number of generations per evolutionary trial being set to 200. Twenty trials were conducted.

The results of two representative trials are indicated in Figure 5-12. The results were fairly consistent across all 20 trials. In each evolution, there was an initial tendency for mutual defection—that is, selfish behavior—but after a few generations (e.g., five to 10), machines evolved that would cooperate if they "recognized" other cooperative machines. These machines received an average of nearly three points for each encounter with similar machines yet

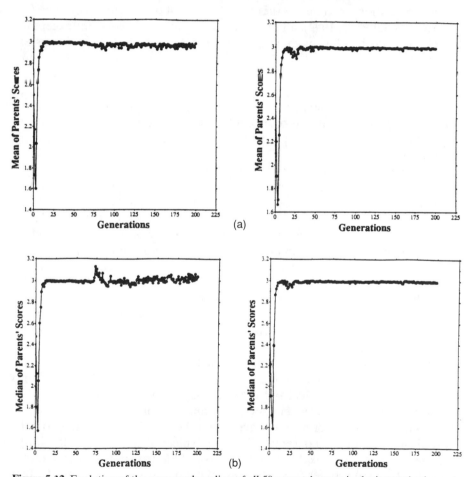

Figure 5-12 Evolution of the mean and median of all 50 parents' scores in the iterated prisoner's dilemma using the payoff function indicated in Table 5-2. (a) Trial 3. (b) Trial 6. The results were similar across all 20 trials.

still retained their initial ability to defect against mutual defectors. As off-spring were created from these new versatile machines, they tended to propagate and dominate the population until it became nearly impossible for even a small number of defectors to persist. The mean population fitness approached but never quite reached 3.0 as each machine tested its potential adversary, apparently checking if it could be exploited by defecting. These results are similar to those reported in Axelrod (1987) and Fogel (1991).

A second set of experiments was designed to examine the rate at which cooperative behavior is achieved in populations of various sizes. Trials were conducted with populations of 50, 100, 250, 500, and 1,000 parents. The execution time increased as the square of the population size due to the round-robin competitions. Although other mechanisms of competition could be formulated (Fogel et al., 1992), all the parameters were held constant except for the population size to provide a measure of comparison to the previous experiment. Each trial was halted at the 25th generation. This decision precluded assessing the stability of evolved cooperation but was necessary in light of the available computing facilities. Thirty trials were conducted with populations of 50 and 100 parents. Five trials were conducted with a population of 250 parents. Multiple trials could not be reasonably completed with populations of 500 and 1,000 parents on the available computing machinery (Sun 4/260).

The mean and median scores of the parents at each generation for all population sizes are shown in Figure 5-13. It would be inappropriate to formulate statistical conclusions in light of the small sample size, but it can be stated that the time to evolve cooperation does not necessarily decrease as the population size is decreased. On the average, cooperative behavior is evidenced before the tenth generation. Thirty additional trials were conducted with populations of 50 and 100 parents to test whether there was any difference in the mean performance of all parents after 25 generations. The two-sided Smith-Satterthwaite t-test indicated that there was insufficient evidence to suggest any difference in average performance ($t = 1.17, P > 0.2$). There appears to be very little qualitative difference between the evolved behavior for populations ranging from 50 to 1,000 parents.

An examination of the best FSM in each trial supports this view. Figure 5-14 indicates a best FSM after 25 generations with the number of parents varying over 50, 100, 250, 500, and 1,000. To simplify notation, FSM_x will refer to the best-evolved FSM when using a population of x parents. FSM_{50} and FSM_{100} are typical (in terms of behavior) of other FSMs generated across the 30 trials with 50 and 100 parents. FSM_{250} is typical of those generated in the five trials with 250 parents. All the best-evolved FSMs are initially cooperative. Each exhibits many states in which mutual cooperation (C, C) elicits further cooperation (C). Continued defection in light of previous mutual defection is also a repeated pattern across all depicted FSMs.

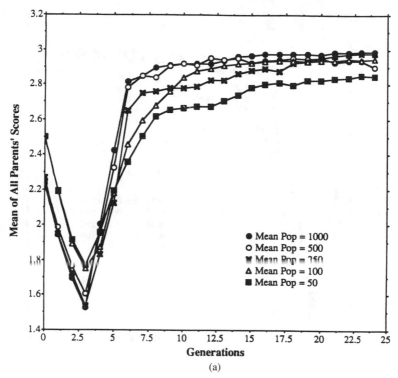

(a)

Figure 5-13 The mean (a) and median (b) of all parents' scores at each generation for populations of 50, 100, 250, 500, and 1,000 parents. The curves for populations of 50 and 100 parents represent the mean of 30 trials. The curve for the population of 250 parents represents the mean of five trials. The curves for populations of 500 and 1,000 are the results of single trials.

It is of interest to compare the behavior that occurs when FSMs from independent trials play each other to that observed when FSMs play copies of themselves. Table 5-3 indicates the results of playing games between the following pairs of FSMs:

(FSM_{50}, FSM_{100})

(FSM_{100}, FSM_{250})

(FSM_{250}, FSM_{500})

(FSM_{500}, FSM_{1000})

(FSM_{50}, FSM_{50})

(FSM_{100}, FSM_{100})

(FSM_{250}, FSM_{250})

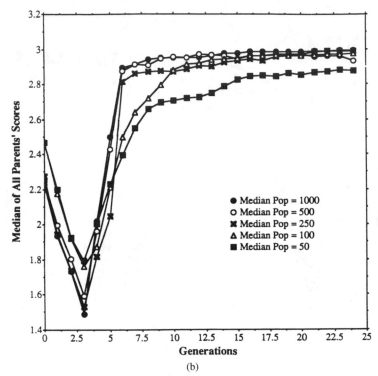

Figure 5-13 *continued*

(FSM_{500}, FSM_{500})

(FSM_{1000}, FSM_{1000}).

Mutual cooperation is rapidly generated when the best FSMs play themselves. This may approximate the actual conditions in each trial, because the final populations of FSMs in any trial may only vary by a relatively small number of mutations. But mutual cooperation is not always the result of encounters between separately evolved FSMs. The games involving (FSM_{50}, FSM_{100}) and (FSM_{100}, FSM_{250}) lead to alternating cooperation and defection $[(C, D), (D, C)]$.

It is speculative but reasonable to presume that each population of FSMs evolves initial sequences of symbols that form patterns that can be recognized by other FSMs. This allows for the identification of machines that will tend to respond to cooperation with cooperation. The specifics of such sequences will vary by trial and may be as simple as merely cooperating on the first move. But the specifics are generally unimportant. When two FSMs from separate trials are played against each other, the resulting behavior may vary widely because no pattern of initial symbols is recognized by either player. Metaphorically,

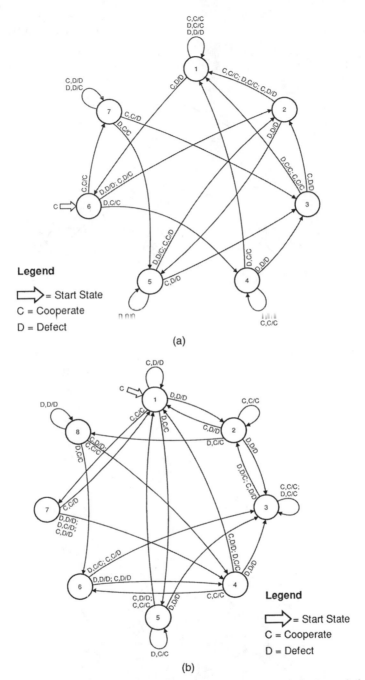

Figure 5-14 The best-evolved finite-state machine after 25 generations for a given population size. Depicted machines for trials with 50, 100, and 250 parents are typical in terms of their behavior. (a) 50 parents. (b) 100 parents. (c) 250 parents. (d) 500 parents. (e) 1,000 parents. Angeline (personal communication) noted an error in Fogel (1993) relating to the output of the machine depicted in (a). The error resulted from an omitted next-state transition. The corrected machine appears above, and the corrected results appear in Table 5-3.

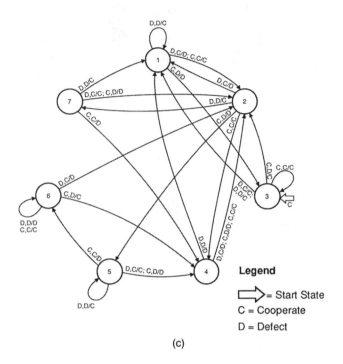

(c)

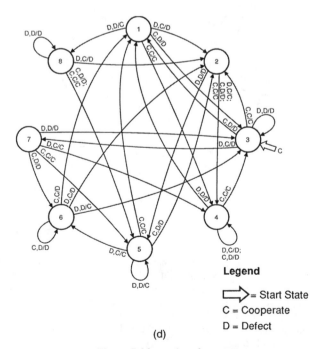

(d)

Figure 5-14 *continued*

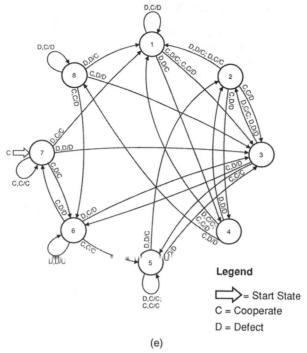

(e)

Figure 5-14 *continued*

they are speaking different languages. Mutual defection or alternating cooperation and defection may be observed even though both FSMs would exhibit mutual cooperation with other FSMs from their respective populations.

A third set of experiments was conducted to examine the effects of varying the reward for defecting in the face of cooperation. When γ_3 becomes sufficiently large, it is expected that mutual cooperation will not evolve (see Eq. 5-1). Two sets of trials were conducted. A series of trials was conducted over 200 generations using 100 parents in which γ_3 ranged over the values 5.25, 5.5, 5.75, and 6.0. Note that when $\gamma_3 = 6.0$, the constraint in Eq. 5-1 is strictly violated. To garner a statistical sense for the variability of evolved behaviors, 10 additional trials were conducted with $\gamma_3 = 6.0$ and executed over 50 generations.

Figure 5-15 indicates the mean of the parents' scores at each generation in each trial with γ_3 ranging over the selected values. All trials exhibit initial mutual defection and, therefore, decreasing payoffs, followed by an increase in performance. Even when $\gamma_3 = 6.0$, the mean of the parents' scores tends toward 3.0 at 200 generations. Figure 5-16 shows the mean of all parents' scores (average of all surviving parents' scores) at each generation when $\gamma_3 = 6.0$ in each of 10 trials. All the depicted curves exhibit similar characteristics, although the parents in trial 3 did not achieve a mean score that would be considered close to 3.0 in light of the tight clustering around this value indicated in the other nine trials.

Table 5-3 The Results of Playing the Iterated Prisoner's Dilemma (a) With Independently Evolved Machines in Which Each Machine Plays Against a Machine That Was Created In a Separate Trial With a Different Number of Parents in the Population and (b) When Each Independently Evolved Best Machine Plays Against Itself. The Population Size Varies Over 50, 100, 250, 500, and 1,000 Parents. Mutual Cooperation, Observed When the Machines Play Themselves, is Not Always The Result of Encounters Between Separately Evolved Machines.

(a)

```
                6  7  3  3  2  1  6  4  4  1  6  4
FSM50           C  C  D  C  D  C  D  C  D  C  D  C              C  D
                                                         →Alternating
FSM100          C  C  D  D  C  D  C  D  C  D  C  D              D  C
                1  7  1  2  8  4  1  1  5  1  5  1
```

```
                1  7  1  5  1  7  1  5  1  2  8  4  1  1  5  1
FSM100          C  C  D  C  C  C  D  C  D  D  C  D  C  D  C  D              C  D
                                                                     →Alternating
FSM250          C  C  D  D  C  D  C  D  C  D  C  D  C  D  C  D              C  D
                3  3  3  2  4  2  4  2  1  1  3  1  3  1  3  1
```

```
                3  3  3  3
FSM250          C  C  C  C          →MUTUAL COOPERATION
FSM500          C  C  C  C
                3  2  4  3
```

```
                3  2  4  3
FSM500          C  C  C  C
                                    →MUTUAL COOPERATION
FSM1,000        C  C  C  C
                7  7  7  7
```

Legend:
C = Cooperate
D = Defect
The numbers above and below the movie indicate the current state for the respective FSM.

(b)

	6	7	3	3	1	1
FSM$_{50}$	C	C	D	C	C	C
FSM$_{50}$	C	C	D	C	C	C
	6	7	3	3	1	1

→MUTUAL COOPERATION

	1	7	1	2	3	2	2
FSM$_{100}$	C	C	D	D	D	C	C
FSM$_{100}$	C	C	D	D	D	C	C
	1	7	1	2	3	2	2

→MUTUAL COOPERATION

	3	3
FSM$_{250}$	C	C
FSM$_{250}$	C	C
	3	3

→MUTUAL COOPERATION

	3	2	4	3
FSM$_{500}$	C	C	C	C
FSM$_{500}$	C	C	C	C
	3	2	4	3

→MUTUAL COOPERATION

	7	7
FSM$_{1,000}$	C	C
FSM$_{1,000}$	C	C
	7	7

→MUTUAL COOPERATION

Legend:
C = Cooperate
D = Defect
The numbers above and below the movie indicate the current state for the respective FSM.

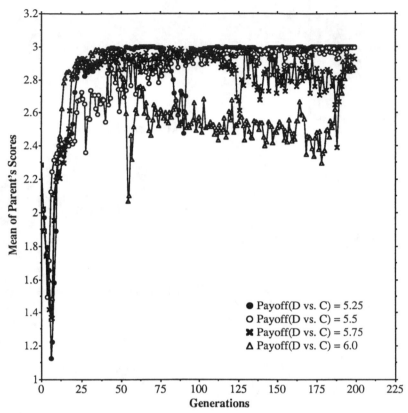

Figure 5-15 The mean of all parents' scores at each generation as γ_3 is increased over 5.25, 5.5, 5.75, and 6.0.

Although it may be tempting to view these data as further evidence of "the evolution of cooperation," this interpretation would not be correct. A careful examination of the best-evolved FSMs after the completion of 50 generations reveals that many of these machines were not cooperative.

Figure 5-17 depicts the highest-scoring FSM at the completion of trial 4 (denoted FSM$_4$). Across all eight states, only two (states 5 and 7) respond to (C, C) with (C), and each of these transitions respond to other states. There is no tendency for endless mutual cooperation with this FSM. But it received a score of 3.003, greater than could be achieved through mutual cooperation alone.

Figure 5-18 indicates the structure of the highest-scoring FSM at the completion of trial 5 (FSM$_5$). The average payoff across all encounters for this machine was 3.537, again greater than could be achieved through mutual cooperation alone. The machine cooperates on the first move. If it meets another

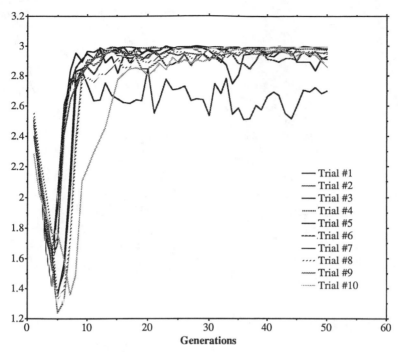

Figure 5-16 The mean of all parents' scores as a function of the number of generations in 10 trials when $\gamma_3 = 6.0$. Trial 3 appears to be a possible outlier.

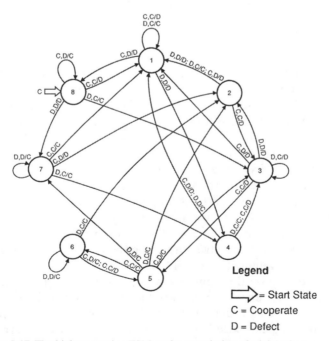

Figure 5-17 The highest-scoring FSM at the completion of trial 4 when $\gamma_3 = 6.0$.

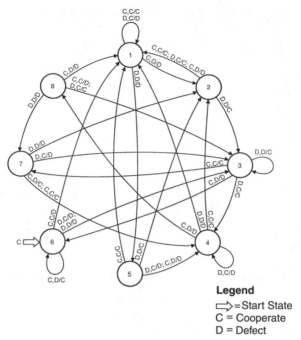

Figure 5-18 The highest-scoring FSM at the completion of trial 5 when $\gamma_3 = 6.0$.

machine that cooperates on the first move, it will defect and transition to state 1. The next two possible input symbols are (D, C) or (D, D). (D, C) elicits further defection; FSM$_5$ will take advantage of the opposition as long as it continues to cooperate. FSM$_5$ would receive a higher score than Tit-for-Tat in these encounters. (D, D) elicits further defection, but not necessarily endless defection, as the machine transitions to state 5 and play then varies.

Both FSM$_4$ and FSM$_5$ are potentially vulnerable. If either were to meet a machine that defected on the first move, each would subsequently cooperate and continue to cooperate as long as the opposing machine defected. The payoff for this behavior would be 0.0. There are two obvious conclusions: (1) Few if any parents in either population (trial 4 or 5) were persistent defectors, and (2) the structural complexity of parent machines was such that simply changing the initial behavior from cooperate to defect was not sufficient to exploit this apparent weakness. The first observation is logically certain; persistent defectors would have caused the mean score for FSM$_4$ or FSM$_5$ to be considerably lower. Neither would have been expected to generate the highest score in their respective population after 50 generations. The second observation can be demonstrated by playing both FSM$_4$

and FSM_5 against copies of themselves with only the starting symbol being varied to defection. Play proceeds as in Table 5-4. When $\gamma_3 = 6.0$, both mutual cooperation and alternating cooperation and defection receive the same reward, an average payoff of 3.0. Simply altering the initial play of cooperation to defection is not sufficient to exploit the potential vulnerability of FSM_4 or FSM_5.

The score of the best-evolved machines across nine out of all 10 trials was greater than 3.0, indicating that the increased benefit for defecting against cooperation had been exploited. Only the best machine in trial 3 (Figure 5-19) received a score of less than 3.0 (2.888). This machine cooperates on the first move and continually cooperates with other machines that reciprocate cooperation. But state 2 of FSM_3 is an absorbing defection state. If the state is entered, the machine will defect forever. Defection is common across the best-evolved machines.

Each curve depicted in Figure 5-16 indicates that the mean score for all parents in each trial was at or below 3.0. The score of the best parent in each trial, however, was typically greater than 3.0. It may be speculated that the best parent machines were not only successful in achieving mutual cooperation with other parents, and in exploiting foolish cooperators, but they may have also evolved a propensity to generate offspring that would tend to cooperate blindly and to bolster their parents' existence further.

Evidence in support of this speculation has been obtained by recording the life span of each parent (i.e., the number of generations that each parent survived). The conditions for trial 1 were repeated and extended over 100 gen-

Table 5-4 The Course of Play When The Best FSM in Trials 4 and 5 Play Against Themselves With Only the Starting Symbol Being Varied to Defection. The Symbol in Parentheses, (C) or (D), Indicates the Initial Play. The Numbers Above and Below The Sequence of Moves Indicate the Current State of the Machine.

	8	8	1	4	3	1	1	8	3	1	1
FSM_4 (C)	C	C	D	D	C	D	C	D	C	D	C

→Alternating C D / D C

FSM_4 (D)	D	C	D	C	D	C	D	C	D	C	D
	8	3	5	7	2	1	8	3	1	1	8

	6	6	6	1	1	5	2	1
FSM_5 (C)	C	C	C	D	D	D	C	C

→Mutual Cooperation

FSM_5(D)	D	D	C	C	D	D	C	C
	6	3	4	2	1	5	2	1

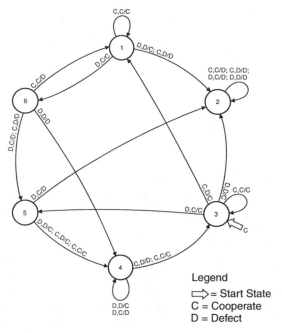

Figure 5-19 The highest-scoring FSM at the completion of trial 3 when $\gamma_3 = 6.0$.

erations. Figure 5-20 indicates the distributions of the life span of each parent at selected generations. The histograms verify steadily increasing parental life spans; that is, in many cases, offspring were unable to displace their parents. This suggests the strong possibility that the parents were generating sacrificial offspring, FSMs that only served to increase the fitness of their parents. The evolutionary simulation apparently generated behavioral castes. Furthermore, despite selection acting solely on individuals with respect to their own success, and despite the initial evolution of purely selfish behaviors, the mean performance of the group (i.e., the reproducing population) ultimately increased to nearly 3.0.

These experiments hint at the ability of evolutionary computation to perform general problem solving. Finite-state machines can in principle generate the behavior of any digital computer. Pierce (1980, p. 62) even suggested that humans may be viewed as finite-state machines. Further, no specific heuristics were added to facilitate modeling the environment. And the environment in question was a dynamic function of the evolution of other finite-state machines. The behavior of the population tended toward that which yielded the maximum possible mean payoff, indicating that the evolutionary computation had gained an understanding of the environment in light of the varying payoffs for alternative behaviors.

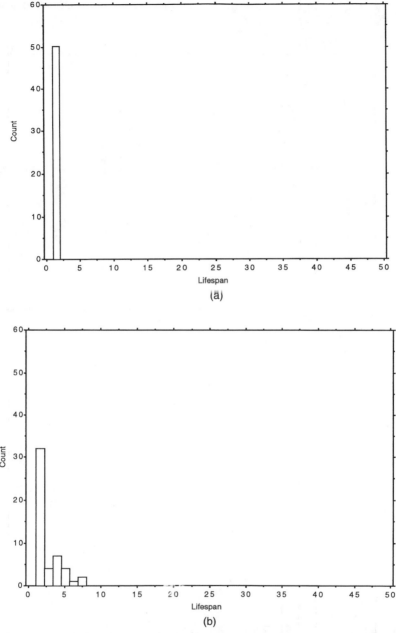

Figure 5-20 Histograms of the life spans (the number of generations that an FSM has survived in the population) of all parents in trial 1 depicted at selected generations. (a) First generation. (b) 25th generation. (c) 50th generation. (d) 75th generation. (e) 100th generation. As the generation number increases, the life span of the parents tends to increase as well. Several of the parent FSMs at the 100th generation were created before the 80th generation. The increasing life span of these parents suggests that they may have created offspring that they could subsequently exploit in competition.

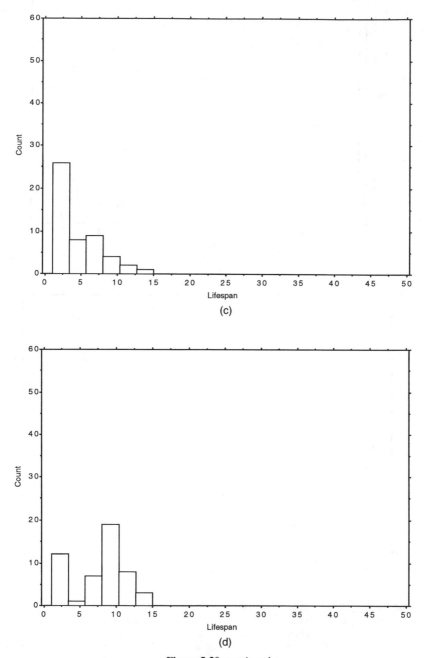

Figure 5-20 *continued*

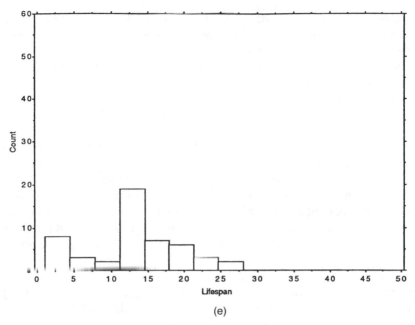

(e)

Figure 5-20 *continued*

5.4 LEARNING HOW TO PLAY CHECKERS
WITHOUT RELYING ON EXPERT KNOWLEDGE

The example offered in section 5.2 demonstrated that evolutionary computation can be used to generate appropriate behavior when facing a competent opponent in the game of tic-tac-toe. The evolutionary approach was adaptive in the sense that the same program could be used to evolve neural networks that would reflect the purpose assigned in terms of the importance of playing for a win, a loss, or a draw. No tuning was required to achieve this feat, and no human expertise about tic-tac-toe was preprogrammed into the evolved neural strategies. The procedure did, however, rely on the prior existence of a competent set of heuristics that could be used as the opposition. A more significant challenge lies in having an evolutionary algorithm learn competent strategies in a complex setting in the absence of a knowledgeable, handcrafted (i.e., human-designed) opponent.

To address this concern, consider the problem of designing an evolutionary algorithm that improves the strategy of play in the game of checkers (also known as draughts, pronounced "drafts") simply by playing successive games between candidate strategies in a population, selecting those that perform well relative to others in the population, making random variations to those strategies that are selected, and iterating this process. Following the previous experiments using the game of tic-tac-toe, strategies can be represented by neural networks. Before providing the algorithmic details, however, the game will be described here for completeness.

Checkers is traditionally played on an eight by eight board with squares of alternative colors of red and black (see Figure 5-21). There are two players, denoted as "red" and "white" (or "black" and "white," but here for consistency with a commonly available Web site on the Internet that allows for competitive play between players who log in, the notation will remain with red and white). Each side has 12 pieces (checkers) which begin in the 12 alternating squares of the same color that are closest to that player's side, with the right-most square on the closest row to the player being left open. The red player moves first, and then play alternates between sides. Checkers are allowed to move forward diagonally one square at a time, or, when next to an opposing checker and there is a space available directly behind that opposing checker, by jumping diagonally over an opposing checker. In the latter case, the opposing checker is removed from play (Figure 5-22). If a jump would in turn place the jumping checker in position for another jump, that jump must also be played, and so forth, until no further jumps are available for that piece. Whenever a jump is available, it must be played in preference to a move that does not jump; how-

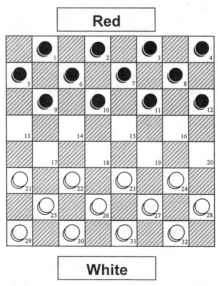

Figure 5-21 The starting position for the game of checkers. Pieces move forward diagonally toward the opposing side of the board. Any piece that reaches the opposing far side becomes a "king" and can thereafter move forward and backward diagonally.

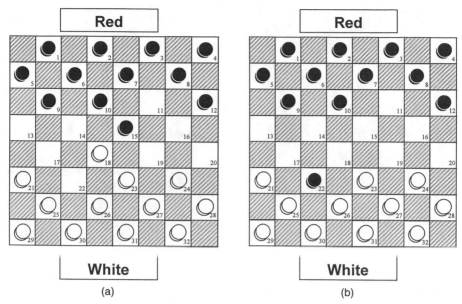

Figure 5-22 An example of a jump move in checkers. Red to move. (a) The red checker on 15 is next to the white checker on 18 and square 22 is open. Red is forced to jump. (b) The result of the jump. The red checker on 15 is moved to 22, and the white piece from 18 is removed from play. Play then switches to white's move, where white must choose between jumping the red piece on 22 by either 25–18 or 26–17.

ever, when multiple jump moves are available, the player has the choice of which jump to conduct, even when one jump offers the removal of more opponents' pieces (e.g., a double jump vs. a single jump). When a checker advances to the last row of the board it becomes a king and can thereafter move diagonally in any direction (i.e., forward or backward). The game ends when a player has no more available moves, which most often occurs by having their last piece removed from the board but can also occur when all existing pieces are trapped (Figure 5-23), resulting in a loss for the player with no remaining moves and a win for the opponent (the object of the game). The game can also end when one side offers a draw and the other accepts.[1]

Unlike tic-tac-toe and many other simpler games, there is no known "value" of the game of checkers. That is, it is not known if the player who

[1] The game can also end in other ways: (1) a player may resign, (2) a draw may be declared when no advancement in position is made in 40 moves by a player who holds an advantage, subject to the discretion of an external third party, and if in match play, (3) a player can be forced to resign if they run out of time, which is usually limited to 60 minutes for the first 30 moves, with an additional 60 minutes being allotted for the next 30 moves, and so forth.

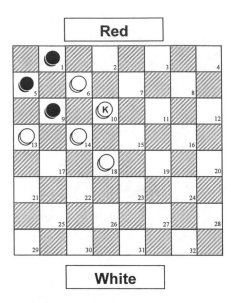

Figure 5-23 Red's move and a losing position for red. Red cannot move any of its pieces. The white checker with the "K" indicates a king. Note that if it were white's move in this situation, white would have three possible moves: 6–2, 10–7, and 18–15. Playing 6–2 would force red to move 1–6, whereupon the white king would jump 10–1 and again red would be unable to move and would therefore lose.

moves first can force a win or a draw. The number of possible combinations of board positions is over 5×10^{20} (Schaeffer, 1996, p. 43), and the game tree of possible sequences of moves remains too large to enumerate. Endgame positions with up to eight pieces remaining on the board have been enumerated and incorporated into some checkers-playing computer programs as lookup tables to determine exactly which moves are best (as well as the ultimate outcome) under these conditions (e.g., in the program *Chinook,* Schaeffer et al., 1996). The number of positions with up to eight pieces is about 440 billion. The number of positions rapidly increases with the number of pieces as a combinatorial function, making an exhaustive listing of longer endgame sequences impractical.

Chellapilla and Fogel (1999) adopted the following protocol for evolving strategies in the game of checkers. Each board was represented by a vector of length 32, with each component corresponding to an available position on the board. Components in the vector could take on elements from $\{-K, -1, 0, +1, +K\}$, where K was the value assigned for a king, 1 was the value for a regular checker, and 0 represented an empty square. The sign of the value indicated whether or not the piece in question belonged to the player (positive) or the opponent (negative). A player's move was determined by evaluating the presumed quality of potential future positions. This evaluation function was structured as a fully connected feed forward neural network with an input layer, two hidden layers, and an output node. The nonlinearity function at each hidden and output node was chosen to be the hyperbolic tangent (tanh, bounded

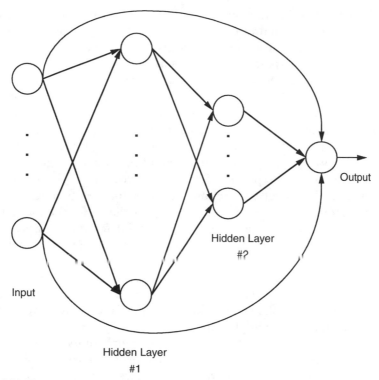

Figure 5-24 The neural network topology chosen for the evolutionary checkers experiments. The networks have 32 input nodes corresponding to the 32 possible positions on the board. The two hidden layers comprise 40 and 10 hidden nodes, respectively. All input nodes are connected directly to the output node with a weight of 1.0. Bias terms affect each hidden and output node but are not shown.

by ± 1) with a variable bias term, although other sigmoidal functions could undoubtedly have been chosen. In addition, all input nodes were connected directly to the output node. Figure 5-24 shows the general structure of the network. At each generation, a player was defined by their associated neural network in which all of the connection weights (and biases) were evolvable, as well as their evolvable king value. For all experiments offered here, each network comprised 40 nodes in the first hidden layer and 10 nodes in the second layer.[2]

[2] These values were chosen after initial experiments with 10 and 8 nodes in each hidden layer gave modestly encouraging results and no further tuning of the number of nodes was undertaken. No claim of optimality is offered for the design chosen. Indeed, the result that reasonable levels of play can be achieved without tuning the neural structure is one of the main points to be made here.

With only one exception, no attempt was made to offer useful features as inputs to a player's neural network. The common approach to designing superior game-playing programs is to perform exactly this sort of intervention where a human expert delineates a series of board patterns or general features that are weighted in importance, positively or negatively (Schaeffer et al., 1992, 1996; Griffith and Lynch, 1997; and others). In addition, entire opening sequences from games played by grand masters and lookup tables of endgame positions can also be stored in memory and retrieved when appropriate. This is exactly the opposite of the approach adopted in Chellapilla and Fogel (1999). The only feature that could be claimed to have been offered is a function of the piece differential between a player and its opponent, owing to the direct connections between the inputs and the output node. The output essentially sums all the inputs that offer the piece advantage or disadvantage. But this is not true in general, for when kings are present on the board, the value K or $-K$ is used in the summation, and as described below, this value is evolvable rather than prescribed by the programmers *a priori*. Thus, the evolutionary algorithm has the potential to override the piece differential and invent a new feature in its place. Absolutely no other explicit or implicit features of the board beyond the location of each piece were implemented.

When a board was presented to a neural network for evaluation, the output node designated a scalar value that was interpreted as the worth of that board from the position of the player whose pieces were denoted by positive values. The closer the output value was to 1.0, the better the evaluation of the corresponding input board. Similarly, the closer the output was to -1.0, the worse the board. All positions that were wins for the player (e.g., no remaining opposing pieces) were assigned the value of exactly 1.0; similarly, all positions that were losses were assigned the value of exactly -1.0.

To begin the evolutionary algorithm, a population of 15 strategies (neural networks), $P_i, i = 1, \ldots, 15$, defined by the weights and biases for each neural network and the strategy's associated value of K, were created at random. Weights and biases were generated by sampling from a uniform distribution over $[-0.2, 0.2]$, with the value of K set initially to 2.0. Each strategy had an associated self-adaptive parameter vector $\sigma_i, i = 1, \ldots, 15$, where each component corresponded to a weight or bias and served to control the step size of the search for new mutated parameters of the neural network. To be consistent with the range of initialization, the self-adaptive parameters for weights and biases were initially set to 0.05.

Each parent generated an offspring strategy by varying all of the associated weights and biases, and possibly the value of K as well. Specifically, for each parent $P_i, i = 1, \ldots, 15$ an offspring $P'_i, i = 1, \ldots, 15$, was created by:

$$\sigma_i'(j) = \sigma_i(j) \times \exp(\tau N_j(0,1)), j = 1, \ldots, N_w$$

$$w_i'(j) = w_i(j) + \sigma_i(j)N_j(0,1), j = 1, \ldots, N_w$$

where N_w is the number of weights and biases in the neural network (here this is 1741), $\tau = \left(\sqrt{2\sqrt{N_w}}\right)^{-1} = 0.1095$, and $N_j(0,1)$ is a standard Gaussian random variable resampled for every j. The offspring king value K' was obtained by:

$$K_i' = K_i + 0.1U_i$$

where U_i was an integer sampled uniformly from $\{-1, 0, 1\}$. Thus, the off-spring's king value had the possibility of incrementing or decrementing by a factor of 0.1, or remaining the same, each with equal likelihood. For convenience, the value of K' was constrained to lie in the range from $[1.0, 3.0]$.

All parents and their offspring competed for survival by playing games of checkers and receiving points for their resulting play. Each player in turn played one checkers game against each of five randomly selected opponents from the population (with replacement). In each of these five games, the player always played red, whereas the randomly selected opponent always played white. In each game, the player scored $-2, 0$, or $+1$ points depending on whether it lost, drew, or won the game, respectively. (A draw was declared after 100 moves for each side.) Similarly, each of the opponents also scored $-2, 0$, or $+1$ points depending on the outcome. These values were somewhat arbitrary but reflected a generally reasonable protocol of having a loss be twice as costly as a win was beneficial. In total, there were 150 games per generation, with each strategy participating in an average of 10 games. After all games were complete, the 15 strategies that received the greatest total points were retained as parents for the next generation, and the process was iterated.

Each game was played using a minimax search of the associated game tree for each board position looking a selected number of moves into the future. For a given board position, all possible moves were enumerated, followed by all of the opponent's possible responses to each possible move, and so forth, up to a preset maximum tree depth, d. By convention, each player's move is termed a "ply"; thus, a move and the opponent's reply consist of two ply. The minimax move for a given ply is determined by selecting the available move that allows the opponent to do the least damage as determined by the evaluation function on the resulting position. For the experiments in Chellapilla and Fogel (1999), d was chosen to be 4 to allow for reasonable execution times (100 generations on a 400-MHz Pentium II required seven days, although no serious attempt was made to optimize the run-time performance of the algorithm). In addition, the ply depth was extended one ply for each forced move that occurred within the first d ply (let f be the number of forced moves in the first d ply) because in these situations the player has no real decision to make. This made it possible for the search tree to end at an odd ply (corresponding to the player's own future move on the $(d + f)$th ply without considering the opponent's response). The best move to make was chosen by iteratively mini-mizing or maximizing over the leaves of the game tree at each ply according to whether or not that ply corresponded to the opponent's move or the player's move.[3] For more on the mechanics of minimax search, see Kaindl (1990).

This evolutionary process, starting from completely randomly generated neural network strategies, was iterated for 100 generations. The best-scoring network at generation 100 was then tested against the authors of the program (Chellapilla and Fogel) using a depth of $d = 6$ (which caused the minimax search for each move to take as long as 60 to 90 seconds but was more typi-cally completed in about 30 seconds). Both authors are novice checkers play-ers, and the program easily defeated them.

The neural network was then used to play against human opponents on an Internet gaming site (www.zone.net). Each player logging on to this site is ini-tially given a rating, R_0, of 1600, and a player's rating changes according to the following formula (which follows the rating system of the United States Chess Federation (USCF)):

$$R_{New} = R_{Old} + C(Outcome - W)$$

where

$$W = \cfrac{1}{1 + 10\left(\cfrac{R_{Opp} - R_{Old}}{400}\right)}$$

Outcome $\in \{1$ if Win, 0.5 if Draw, 0 if loss$\}$

and for ratings less than 2100, $C = 32$.[4]

---

[3] When evaluating a board position that would result from the player's move, the signs of all of the inputs as well as the output were flipped, with the selection being performed to find the move that maximized the output. When evaluating a board that would result from an opponent's move, the signs and output remained as initially stated above, with the selection being performed to find the move that minimized the output (thereby assuming that the opponent would pick the move that did maximum damage). The procedure to flip the signs of inputs and outputs is unnec-essary and was removed in subsequent efforts that used an alpha-beta search to accelerate the minimax procedure. These efforts are not described here. Moreover, note that an asymmetry was introduced by flipping the input signs, in that a neural network will not generally be an odd func-tion (defined as $f(-x) = -f(x)$). The effect that this characteristic had on the learning ability of the evolutionary algorithm is unknown.

[4] More complicated transformations are applied for ratings that switch between designated classes above 2100 points, and the value for C changes as well. These situations were not relevant to the scores attained here. The formulas above pertain legitimately to players with established ratings based on 20 or more games, but the Internet gaming zone appeared to use this formula consistently. The USCF uses a different rating formula for players with under 20 games. In essence, the Internet gaming zone estimates the player's performance of their first 20 games to be 1600.

Over the course of a month, 100 games were played against opponents on this Web site. Games were played until (1) a win was achieved by either side, (2) the human opponent resigned, or (3) a draw was offered by the opponent and (i) the piece differential of the game did not favor the neural network by more than one piece and (ii) there was no way for the neural network to achieve a win that was obvious to the authors, in which case the draw was accepted. There was a fourth condition which occurred infrequently in which the human opponents abandoned the game without resigning (by closing their graphical-user interface), thereby leaving their own rating intact. The Internet gaming zone decremented 100 points from a player's rating for every 10th time they abandoned a game, but this did not appear to be a sufficient deterrent for some people. When an opponent abandoned a game in competition with the neural network, a win was counted if the neural network had an obvious winning position (one where a win could be forced easily in the opinion of the authors) or if the neural network was ahead by two or more pieces. Otherwise, the game was not recorded. (This occurred one time and was probably the result of a faulty modem connection for the human opponent.) In no cases were the opponents told that they were playing a computer program, and no opponent ever commented that they believed their opponent was a computer algorithm.

Opponents were chosen based primarily on their availability to play (i.e., they were not actively playing someone else at the time) and to ensure that the neural network competed against players with a wide variety of skill levels. In addition, an attempt was made to balance the number of games played as red or white. In all, 44 games were played as red. Figure 5-25 shows a histogram of the number of games played against players of various ratings along with the win-loss-draw record attained in each category. The evolved neural network performed well against players rated 1700 and lower, and had almost as many losses as wins against opponents rated between 1700 and 1800. In contrast, it earned no wins (and three draws) against opponents rated over 1900. Figure 5-26 shows the sequential rating of the neural network and the rating of the opponents played over all 100 games. Table 5-5 provides a listing of the class intervals and designations accepted by the USCF. The highest rating attained by the evolved neural network was 1825.4 on game 85.

The final rating of the neural network was 1750.8, which places it subjectively as a better than median Class B player. For comparison, the top 10 rated players registered at this Internet site (as of December 13, 1998) had ratings of:

1. 2308
2. 2294
3. 2272
4. 2248

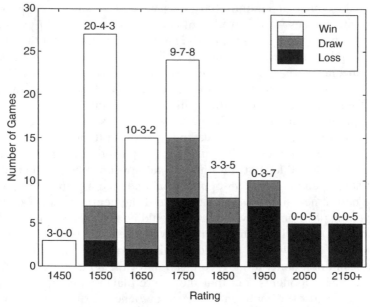

Figure 5-25 The performance of the evolved neural network after 100 generations, played over 100 games against human opponents on an Internet checkers site. The histogram indicates the rating of the opponent and the associated performance against opponents with that rating. Ratings are binned in intervals of 100 units (i.e., 1650 corresponds to opponents who were rated between 1600 and 1700). The numbers above each bar indicate the number of wins, draws, and losses, respectively. Note that the evolved network generally defeated opponents who were rated less than 1700 and played to about an equal number of wins and losses with those who were rated between 1700 and 1800. No wins were obtained against opponents rated 1900 or higher.

5. 2246

6. 2239

7. 2226

8. 2224

9. 2221

10. 2217

Thus, the top 10 players were all at the master level.

The best performance of the evolved network was likely recorded in a game against a player rated 1926 (Class A), which ended in a draw. The sequence of moves proceeded as follows. Certain moves are annotated, but note that these annotations are not offered by an expert checkers player (instead

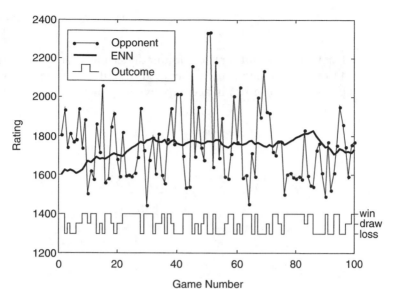

Figure 5-26 The sequential rating of the evolved neural network (ENN) over the 100 games played against human opponents. The graph indicates both the network's rating and the corresponding rating of the opponent on each game, along with the result (win, draw, or loss). The highest rating for the ENN was 1825.4 on game 85. The final rating after game 100 was 1750.8, placing the ENN as a better than median Class B player.

Table 5-5 The Relevant Categories of Player Indicated by the Corresponding Range of Rating Score.

Class	Rating
Senior Master	2400+
Master	2200–2399
Expert	2000–2199
Class A	1800–1999
Class B	1600–1799
Class C	1400–1599
Class D	1200–1399
Class E	1000–1199
Class F	800–999
Class G	600–799
Class H	400–599
Class I	200–399
Class J	below 200

223

being offered here by the author). Undoubtedly, a more advanced player might have different comments to make at different stages in the game. Selected positions are also shown in accompanying figures.

```
Game Against Human Rated 1926
Human Plays Red, Computer Plays White
(f) denotes a forced move
Comments on moves are offered in brackets
```

Red's Move	White's Move	Time Required (sec)	Board Evaluations
1.R:11-16	1.W:23-19	11.38	103720
2.R:16-23(f)	2.W:26-19	5.08	45435

[early swap]

3.R:8-11	3.W:19-15	25.55	234312

[double swap]

4.R:11-18	4.W:22-15(f)	0	0
5.R:10-19(f)	5.W:24-15(f)	0	0
6.R:7-10	6.W:27-24	27.92	258435

[move to swap]

7.R:10-19(f)	7.W:24-15(f)	0	0
8.R:6-10			

[red swaps]

	8.W:15-6(f)	0	0
9.R:1-10(f)	9.W:25-22	34.03	310186
10.R:9-14	10.W:30-26	45.84	421403
11.R:3-7	11.W:22-17	21.74	199619

[trying for a king?]

12.R:4-8	12.W:26-23	17.13	157548
13.R:8-11	13.W:28-24	15.07	139413
14.R:11-15			

[moving toward attacking the back rank]

	14.W:32-28	16.23	149198
15.R:7-11	15.W:29-25	10.09	93027
16.R:15-18			

[attacking position 23]

	16.W:23-19	7.58	70239

[move away]

17.R:18-23	17.W:24-20	8.26	76159

[preventing king]

18.R:5-9	18.W:17-13	7.52	68502

[swap?]

19.R:14-18	19.W:13-6(f)	0	0

[here's the swap]

20.R:2-9(f)	20.W:21-17	7.12	64641
21.R:9-13			

[threatening 17]

	21.W:19-15	2.29	21212

[sacrificing]

22.R:10-19	

[take the piece on 15; frees piece on 17. Was this a mistake? Should have double jumped to get king 13-22-29? See Figure 5-27]

	22.W:17-14	1.82	16879
23.R:13-17	23.W:25-21	5.31	49061

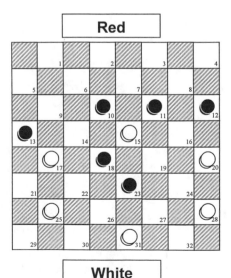

Figure 5-27 Red (human) to move. Red has the advantage and must choose between the jump 10-19 and the double jump 13–22–29. Red chooses 10–19, which appears to have been a mistake.

[saves piece]

24.R:17-22

[red to get king]

```
                24.W:14-9      5.24              48470
25.R:11-15      25.W:9-6       3.93              36253
26.R:22-26
```

[swap to free up kings]

```
                26.W:31-22(f)  0                 0
27.R:18-25(f)   27.W:6-1       3.99              35983
```

[king]

```
28.R:15-18      28.W:1-6       5.17              47804
29.R:23-27      29.W:6-10      13.8              127411
30.R:18-22
```

[trying to advance as many pieces as possible]

```
                30.W:10-15     11.41             105782
```

[good move because the obvious 19-23 to move away leads to a double
jump. See Figure 5-28]

```
31.R:19-23      31.W:20-16     7.53              69727
32.R:12-19(f)   32.W:15-24-31(f) 0               0
33.R:25-29
```

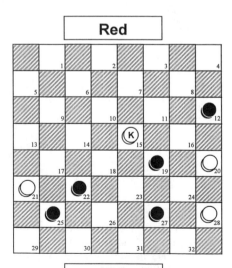

Figure 5-28 The position after white (computer) moves 10–15, threatening red's piece on 19. The obvious move for red is 19–23 to flee, but this leads to white's 20–16, setting up a double jump for white after sacrificing the checker on 16.

[king]

```
                  33.W:21-17      0.36              3340
34.R:22-25        34.W:17-13      0.42              3849
35.R:25-30        35.W:13-9       0.53              4977
```

[racing for king]

```
36.R:30-25        36.W:9-5        0.81              7535
37.R:23-26
```

[exchanging a piece for a king]

```
                  37.W:31-22(f)   0                 0
38.R:25-18(f)     38.W:5-1        0.34              3154
```

[king]

```
39.R:29-25        39.W:1-5        1.75              10005
40.R:18-14        40.W:28-24      0.94              8609
```

[finally]

```
41.R:25-22        41.W:24-19      1.45              13403
42.R:22-18        42.W:19-16      0.67              6228
43.R:18-15
```

[in pursuit]

```
                  43.W:16-12      0.76              6973
44.R:15-11
```

[trapped the white piece. See Figure 5-29]

```
                  44.W:5-1        0.32              2947
45.R:14-9         45.W:1-5        0.1               920
46.R:9-6          46.W:5-1        0.07              618
47.R:6-9          47.W:1-5        0.1               920
48.R:9-6          48.W:5-1        0.07              618
49.R:6-9          49.W:1-5        0.1               920
50.R:9-6          50.W:5-1        0.07              618
```

[how long will they continue this?]

```
51.R:6-9          51.W:1-5        0.1               920
52.R:9-6          52.W:5-1        0.07              618
53.R:6-9          53.W:1-5        0.1               920
54.R:9-6          54.W:5-1        0.07              618
55.R:6-9          55.W:1-5        0.1               920
56.R:9-6          56.W:5-1        0.07              618
57.R:6-9
```

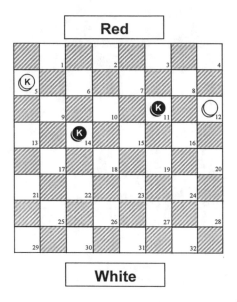

Figure 5-29 The end of the game. Red (human) has moved 15–11, pinning white's piece on 12. But white's king on 5 cannot then be pinned, leading to draw.

```
[Red offers draw]

          57.W:1-5        0.1                    920

[White accepts the draw]
```

In retrospect, the 1926-rated player made perhaps two errors at moves 22 and 30.

Two other noteworthy games were played for wins against upper-level Class B players. In the first, the human was rated 1771 and played white. The evolved neural network finished the game with a flourish.

```
Game Against Human Rated 1771
Human Plays White, Computer Plays Red
(f) denotes a forced move
Comments on moves are offered in brackets
```

White's Move	Red's Move	Time Required (sec)	Board Evaluations
	0. 9-13	10.43	94371

```
[The neural network's standard opening — is it weak?]

1.W:22-18      1.R:11-15      18.09                   163600

[early exchange]

2.W:18-11(f)   2.R:7-16       5.21                    47031
```

[the neural network likes to head to the side]

3.W:25-22	3.R:5-9	34.45	315582
4.W:22-18	4.R:3-7	30.81	282594
5.W:29-25	5.R:1-5	16.01	146615
6.W:25-22	6.R:16-19	10.49	95491

[double exchange]

7.W:23-16	7.R:12-19(f)	0	0
8.W:24-15(f)	8.R:10-19(f)	0	0
9.W:27-24			

[attacking position 19]

9.R:7-11	15.67	143903

[sacrificing the piece on 10 ... setting up something?]

10.W:24-15(f)	10.R:9-14	11.31	102969

[trading for position on a double jump]

11.W:18-9(f)	11.R:11-18-25	4.01	36803

[double jump, with a king to follow]

12.W:26-23	12.R:5-14(f)	0	0

[take back the trade — overall +1 piece]

13.W:23-19	13.R:25-29	25.64	235142

[king]

14.W:31-26	14.R:6-10	20.77	190334
15.W:19-16	15.R:8-12	14.08	129084
16.W:16-11			

[simultaneous threat to two back rank pieces]

	16.R:12-16	17.72	163200
17.W:28-24	17.R:29-25	23.34	214543
18.W:32-27	18.R:16-20	17.10	157150
19.W:24-19	19.R:13-17	17.11	156196
20.W:26-23	20.R:25-22	9.74	89193
21.W:19-16	21.R:22-26	10.03	91518
22.W:23-19	22.R:26-31	5.35	48803

[attacking position 27]

23.W:27-23   23.R:17-22   11.32   103881

[going for king]

24.W:11-7   24.R:2-11(f)   0   0
25.W:16-7(f)   25.R:31-27   12.59   115134

[attacking 23]

26.W:7-2

[gave up on 23, got king]

                  26.R:27-18(f)   0   0
27.W:19-16   27.R:18-23   10.92   100479
28.W:16-11   28.R:22-26   14.42   133235

[going for king]

29.W:2-6

[probably a mistake here for white. See Figure 5-30]

                  29.R:4-8   6.76   62405

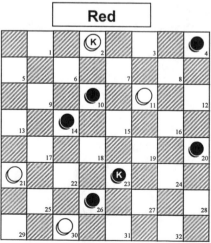

Figure 5-30 White (human) chooses to move their king 2–6. This is probably a mistake because it sets up a series of forced moves where red (computer) manipulates white into a double jump that effectively wins the game.

[giving up on 10, and why make this move which gives white a king??
We'll see later]

30.W:11-4

[graciously accepting the present]

                30.R:20-24   5.16   45039

[moving for king]

31.W:6-15(f)   31.R:23-27   4.90   43937

[set up for a double jump after the sacrifice; note that the 11-4 jump
was necessary to open up the double jump]

32.W:30-23(f) 32.R:27-18-11(f)   0   0

[game over with this jump. See Figure 5-31]

33.W:21-17   33.R:14-21(f) 0   0
34.W:4-8(f)   34.R:11-4(f)   0   0

[game over] red (computer) wins

The other game was played against a player rated 1757. The game is noteworthy because the evolved neural network was able to complete an endgame to a positive conclusion, rather than be content with a draw.

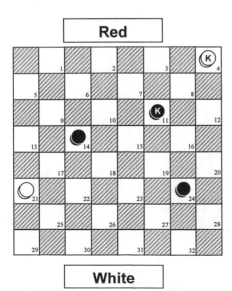

Figure 5-31 White (human) to move and is in an obvious losing situation. Both potential moves will result in immediate jumps. The game is over and the computer wins.

Game Against Human Rated 1757
Human Plays White, Computer Plays Red
(f) denotes a forced move
Comments on moves are offered in brackets

White's Move	Red's Move	Time Required (sec)	Board Evaluations
	0.R:9-13	10.39	94371

[standard opening]

1.W:23-18	1.R:6-9	15.72	140918
2.W:18-15			

[swapping early for the human]

	2.R:10-19	5.80	53813
3.W:24-15(f)	3.R:11-18(f)	0	0
4.W:22-15(f)	4.R:13-17	27.41	250355

[swapping more]

5.W:21-14(f)	5.R:9-18(f)	0	0
6.W:25-21			

[why not 26-23 for a free piece? As game follows we see that the piece
is not free after all]

	6.R:7-11	33.20	306330
7.W:26-23			

[now human plays this move]

	7.R:1-6	38.47	354113
8.W:23-14(f)	8.R:11-18(f)	0	0

[an exchange]

9.W:31-26	9.R:6-10	30.63	280847

[swapping]

10.W:14-7(f)	10.R:3-10	4.65	41587
11.W:29-25	11.R:8-11	15.95	146561
12.W:26-23			

[threatened double jump]

	12.R:10-14	15.98	147427

[blocked]

```
13.W:23-19      13.R:11-16      18.41                     168417
14.W:19-15
```

[slipping past the attack]

```
                14.R:4-8        20.77                     190870
15.W:15-10      15.R:16-20      13.73                     126243
```

[again, the neural network's tendency to move to the side]

```
16.W:30-26      16.R:8-11       10.18                     93194
```

[gives white a king]

```
17.W:10-7
```

[moving for king]

```
                17.R:11-15      9.95                      91453
```

```
18.W:7-3
```

[king]

```
                18.R:12-16      13.31                     122033
19.W:3-8        19.R:2-6        23.90                     219420
20.W:8-12
```

[threatened jump but doesn't consider the possible 12-19 reply after
6-10. See Figure 5-32]

```
                20.R:6-10       5.88                      53798
21.W:12-19(f)   21.R:15-24-31(f)0                         0
```

[. . . and king! with threat to jump]

```
22.W:26-22      22.R:18-23      2.34                      21435
```

[and another king]

```
23.W:22-17      23.R:14-18      3.93                      36115
24.W:17-14
```

[swapping to get a king]

```
                24.R:10-17(f)   0                         0
25.W:21-14(f)   25.R:23-26      2.23                      20136
```

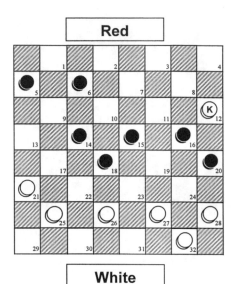

Figure 5-32 White (human) has moved their king 8–12, attacking the red checker at 16. But red (computer) will move 6–10, setting up a double jump to king.

```
[going for king]

26.W:14-10

[going for king]

                26.R:18-23      2.26                20862
27.W:10-7       27.R:26-30      2.50                23221

[king]

28.W:25-21      28.R:30-25      5.22                48385
29.W:21-17

[running away]

                29.R:25-21     11.44               105541
30.W:17-14      30.R:21-25      6.85                63411

[gave up?]

31.W:7-2

[king]

                31.R:31-26     13.13               121482
32.W:14-10      32.R:26-22     12.73               118150
```

```
33.W:10-7      33.R:22-17      12.65                116774
34.W:7-3
```

[king]

```
               34.R:17-13      34.92                320376
35.W:3-7       35.R:13-17      23.85                220616
```

[toggled last position]

```
36.W:7-10      36.R:17-13      39.70                368603
```

[again toggled last position]

```
37.W:10-14     37.R:23-26      17.58                162416
```

[go get the other king]

```
38.W:14-18     38.R:26-30      26.28                242204
```

[king]

```
39.W:18-15     39.R:25-21      29.57                272184
40.W:15-10     40.R:30-26      21.90                201777
41.W:10-15
```

[white's in no hurry]

```
               41.R:21-17      38.49                353823
42.W:15-11     42.R:13-9       42.42                391092
43.W:2-7       43.R:9-6        66.46                613561
```

[closing in]

```
44.W:11-15     44.R:26-22      83.96                760957
```

[moves are taking a long time to find, nearly 1.5 minutes]

```
45.W:15-11     45.R:6-2        65.41                599223
46.W:7-10      46.R:17-21      39.84                369091
```

[backed away, why?]

```
47.W:11-15     47.R:21-25      24.27                225272
```

[where is he going?]

```
48.W:15-19     48.R:22-17      40.58                376237
49.W:19-23
```

[maybe trying to free up his pieces]

```
                 49.R:17-13     49.55              459028
50.W:23-26       50.R:25-22     25.85              238989
```

[moved king to swap]

```
51.W:26-17(f)    51.R:13-22(f)  0                  0
52.W:10-15       52.R:2-6       7.84               71957
53.W:15-11       53.R:22-26     15.38              141419
```

[not very aggressive]

```
54.W:11-15       54.R:26-23     8.46               77746
55.W:15-11
```

[forced into a bad spot here]

```
                 55.R:20-24     8.03               73406
```

[sets up double jump]

```
56.W:28-19(f)    56.R:23-16-7(f) 0                 0
```

[See Figure 5-33]

```
57.W:32-27       57.R:6-10      0.57               5167
58.W:27-23       58.R:5-9       0.18               1618
59.W:23-19       59.R:7-2       0.28               2682
```

[doesn't have the killer instinct]

```
60.W:19-16       60.R:2-6       0.64               6093
```

[what are you afraid of?]

```
61.W:16-12       61.R:6-1       1.74               16058
62.W:12-8(f)     62.R:10-7      1.25               11713
```

[good]

```
63.W:8-3         63.R:7-11      0.22               2033
```

[game over]

```
64.W:3-8         64.R:11-4(f)   0                  0
```

[takes the piece]

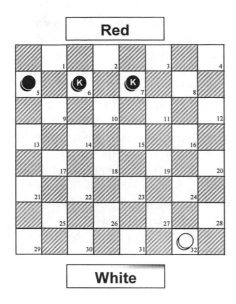

Figure 5-33 The position after red (computer) double jumps 23–16–7, leaving white (human) in a desperate position. Red wins eight moves later.

Overall, the results indicate the ability for an evolutionary algorithm to start with essentially no preprogrammed information in the game of checkers (except the possibility for using piece differential as indicated above) and learn, over successive generations, how to play at a level that is challenging to many humans. The neural network was not able to play at the master level or higher, and this is likely due in part to the limited ply that was employed.

This handicap is particularly evident in the endgame, where it is not uncommon to find pieces separated by several open squares, and a search at $d = 6$ may not allow pieces to effectively "see" that there are other pieces within eventual striking distance. Moreover, the coordinated action of even two pieces moving to pin down a single piece can necessitate a long sequence of moves where it is difficult to ascribe advantage to one position over another until the final result is in view. Finally, it is well known that many endgame sequences in checkers can require very high ply (e.g., 20–60, Schaeffer et al., 1996), and all of these cases were simply unavailable to the neural network to assess. With specially designed computer hardware, it would be possible to implement the best neural network directly on a chip and greatly increase the number of boards that could be evaluated per unit time, and thereby the ply that could be searched. Under the available computing environment, the speed was limited to evaluating approximately 10,000 possible board positions per second. For comparison, Deep Blue was able to evaluate 200 million chess boards per second (Hoan, cited in Clark, 1997).

Another limitation of the procedure was in the use of minimax as a strategy for choosing the best move. Although this is a commonly accepted protocol, it is not always the best choice for maximizing the chances of obtaining a win against an opponent that may make a mistake. By assuming that the opponent will always make the move that is worst from the player's perspective, the player must play conservatively, minimizing that potential damage. This conservatism can work against the player, because when offered the choice between one move that engenders two possible opponent responses, each with values of say +0.05 and +0.2 points, respectively, and another move with two possible responses of 0.0 and +0.9 points, the minimax strategy will favor the first move because it can at worst still yield a gain of +0.05. But the qualitative difference between +0.05 and 0.0 is relatively small (both are effectively even positions). If the second move had been favored there would have been the potential for the opponent to make an error, thereby leaving them in a nearly certain defeat (corresponding to the board evaluated at +0.9). The proper heuristic to use when evaluating the relative advantage of one move over another is not always clear.

To summarize, the information given to the neural networks was essentially limited to:

1. A representation defining the location of each piece (and its type) on the board.

2. A variable coding value for a king.

3. A mechanism for computing all possible legal moves in any potential state of the game.

4. A heuristic for searching ahead up to six ply.

5. A heuristic (minimax) for selecting which move to favor in light of the neural network evaluation function.

6. The potential to use piece differential as a feature.

None of these capabilities is much different from those that novice human players bring to their first game. They are told the rules of how pieces move, thereby giving them the potential to make legal moves. They are told the object of the game, and the most direct manner to achieve that object is to remove the opponent's pieces; therefore, having more pieces than your opponent is an evident subgoal. They are told that kings have different properties than regular pieces, and they must choose some internal representation to separate these two types of pieces. And they are told that the game is played in turns, so it is again clearly evident that moves must be considered in light of what moves the opponent is likely to make in response. The novice human player also recognizes the spatial characteristics of the board, the nearness or distance between pieces, a series of empty squares in a row indicating the potential for moving unimpeded, and other nuances that carry over from recognizing patterns in everyday life. The neural network evolved here had no knowledge of the spatial nature of the game; its board was simply a 32-component vector rather than an eight by eight checkerboard. It would be of interest to assess the performance of neural networks that could evaluate board positions based on such spatial features. Yet, even with this handicap, the evolutionary algorithm was able to learn how to play competent checkers based essentially on the information contained in "win, lose, or draw."

5.5   DISCUSSION

None of the evolutionary simulations described above is in any sense optimum. It is easy to imagine methods for improving each example by incorporating more appropriate mutation distributions or selection mechanisms or representations. But intelligence does not demand optimality. Evolving species and individuals are in competition with themselves, not with any presumed "right answer." There are no absolute measures of intelligence, nor can there be under the definition offered in Chapter 1. Only a comparative evaluation of intelligent behavior can be meaningful. The more successful a system is at evolving behaviors across a wider range of environments with respect to more diverse goals, the more intelligent it is. Intelligence is a property of a system, not a solution.

Although a comparative assessment of intelligence requires a subjective judgment, the methods involved in generating intelligent behavior through evolution can be identified with complete objectivity. The evolutionary algorithms used in all three examples have four fundamental similarities. Each incorporates (1) a stochastic exploration of the state space of possible solutions, (2) a stochastic selection mechanism for purging solutions unworthy of further attention, (3) a mechanism for observing and affecting the environment, and (4) an explicit or implicit mechanism for predicting the future behavior of the environment based on alternative actions. These are the features that ensure versatility, the "essential ingredient of intelligent behavior" (Fogel et al., 1966, p. 3).

The more familiar description of a procedure that incorporates these features is the scientific method. The resemblance between the process of evolution and the process of scientific inquiry is striking. Individual organisms serve as hypotheses concerning the fundamental physics of their environment. Their behavior is an inductive inference concerning anticipated futures, and survival demonstrates an understanding. Offspring carry the "heredity of reasonableness" (Fogel et al., 1966, p. 112) as well as additional information as a result of genetic transformations. Simulating the scientific method and simulating evolution lead to the same basic computational procedure, a general problem-solving technique. But the recognition of intelligence as an evolutionary learning process is hardly new (Turing, 1950; Bremermann, 1958; Booker, 1982; Conrad, 1988; and others), nor has it been only recently demonstrated (Fogel, 1964; Fogel et al., 1966; Reed et al., 1967; Atmar, 1976; and others).

In that regard, the examples offered here should only be considered a contribution to the argument if they serve to highlight the fundamental importance of prediction and control (the allocation of resources) in intelligent behavior. Mere optimization is only of minor importance. To be sure, optimization is intimately associated with evolution (Mayr, 1988, p. 104), but if, for example, optimization is executed simply to discover the extremum of a static topography, then adaptation is made with respect to only a single goal and in only a single environment. The intelligence of such a system is inconsequential. Only when the adaptive topography to be searched is a temporal function, one that is predicted and affected by the evolutionary process itself, should an observer take interest in assessing the system's potential intelligence.

A general problem-solving technique must use representations for solutions that allow for arbitrary transformations from sensed data to elicited action. Finite-state machines and neural networks provide for such universal mappings within the memory limitations of the computer. Symbolic expressions may also be used for such application (Koza, 1992), although no greater generality is achieved. But in each case, care must be taken as to how much a priori information is provided to the learning procedure. It is certainly useful to include operators that are specifically tailored to the problem at hand (e.g., the introduction of trigonometric functions in a spiral classification task; Koza, 1992, pp. 445–457), but their necessity suggests limited robustness. It would seem more prudent to examine cases where system responses are elicited by arbitrary mappings operating on primitive sensory inputs, as described in the neural tic-tac-toe and checkers experiments.

As tic-tac-toe is a common and relatively simple game, the profundity of the success of the evolutionary experiments may be overlooked by a casual observer. Consider a two-player game that incorporates an array of nine elements. Each player places markers in the array. Certain patterns in the array portend wins, losses, and draws, but no a priori information about the characteristics of these patterns is offered. Feedback concerning the quality of play only comes at the end of the game, when the final result becomes known. Or, even more restrictive, feedback may come only after a series of games are played so that the contribution to overall performance from any single game is unknown. Determining an optimal strategy under these conditions appears to be a significant challenge, and these are exactly the conditions that were presented to the evolutionary algorithm. Tic-tac-toe may seem simple because the notion of *in-a-row* is common in human experience and the objective of the game is agreed upon before play begins. If the patterns that define the wins, losses, and draws are arbitrary, the problem becomes much more difficult.[5]

The game of checkers is markedly more complicated than tic-tac-toe. The ability to coevolve competent (albeit not master level) players starting from little more than the most basic knowledge of the game represents a considerable achievement. As with the protocol for evolving tic-tac-toe players, with the exception of piece differential, no explicit hints were given as to what would constitute good or bad moves at any stage of the game. This stands in sharp contrast with the protocol of Samuel (1959) where preselected features were offered to a deterministic algorithm that could update the linear contribution of each feature into an overall board evaluation function. Feedback on the quality of play was again limited to an overall point score that represented the collective worth of a series of games. The result of each individual game was not immediately available as a measure of performance. Further, no handtailored opponent was used as a benchmark against which to evolve; in contrast, the evolutionary process bootstrapped its performance based solely on the strategies available within each current population.

The three simple experiments offered here illustrate evolved intelligent behavior. In each case, starting with completely random solutions to the problem at hand, the computer was able to generate increasingly appropriate behaviors in light of diverse criteria. It adapted its behavior to meet its goals across a range of environments. If intelligent decision making is viewed as a problem of optimally allocating available resources in light of environmental demands and goals, then these experiments do indeed serve as a foundation for the thesis that machine intelligence can be achieved by simulating evolution.

REFERENCES

Akaike, H. (1974). "A New Look at the Statistical Model Identification," *IEEE Trans. Auto. Control,* Vol. AC-19:6, pp. 716–723.

Anderson, B.D.O., R. R. Bitmead, C. R. Johnson, P. V. Kokotovic, R. L. Kosut, I. M. Y. Mareels, L. Praly, and B. D. Riedle (1986). *Stability of Adaptive Systems: Passivity and Averaging Analysis.* Cambridge, MA: MIT Press.

Angeline, P. J. (1994). Personal Communication, Loral Federal Systems, Owego, NY.

---

[5]If I may briefly digress to an example, I have often started lectures over the past couple of years by playing such a game with a member of the audience. I allow the participant to move first on a 3×3 grid, but do not offer what the objective of the game might be. The objective is quite simple: A win is achieved when a player places three markers on the outer squares of the grid. That is, the player moving first can force a win simply by avoiding the middle square. After we complete two or three games, I ask if anyone in the audience has any idea what the object of the game might be. The usual response is laughter. I have carried this out over as many as 10 games, yet no one has ever offered me the correct objective, despite its obvious simplicity.

Atmar, J. W. (1976). "Speculation on the Evolution of Intelligence and Its Possible Realization in Machine Form." Sc.D. diss., New Mexico State University, Las Cruces.

Axelrod, R. (1980a). "Effective Choice in the Prisoner's Dilemma," *J. Conflict Resolution,* Vol. 24, pp. 3–25.

Axelrod, R. (1980b). "More Effective Choice in the Prisoner's Dilemma," *J. Conflict Resolution,* Vol. 24, pp. 379–403.

Axelrod, R. (1984). *The Evolution of Cooperation.* New York: Basic Books.

Axelrod, R. (1987). "The Evolution of Strategies in the Iterated Prisoner's Dilemma." In *Genetic Algorithms and Simulated Annealing,* edited by L. Davis. London: Pitman, pp. 32–41.

Barron, A. R. (1982). "Complexity Approach to Estimating the Order of a Model." Elect. Engineering 378B Final Report, Information Systems Laboratory, Stanford, CA.

Barto, A. G., R. S. Sutton, and C. W. Anderson (1983). "Neuron-like Adaptive Elements That Can Solve Difficult Learning Control Problems," *IEEE Trans. Sys., Man and Cybern.,* Vol. SMC-13:5, pp. 834–846.

Booker, L. (1982). "Intelligent Behavior as Adaptation to the Task Environment." Ph.D. diss., University of Michigan, Ann Arbor.

Boyd, R., and J. P. Lorberbaum (1987). "No Pure Strategy is Evolutionarily Stable in the Repeated Prisoner's Dilemma," *Nature,* Vol. 327, pp. 58–59.

Bremermann, H. J. (1958). "The Evolution of Intelligence. The Nervous System as a Model of Its Environment." Technical Report No. 1, Contract No. 477(17), Dept. of Mathematics, University of Washington, Seattle, July.

Chellapilla, K., and D. B. Fogel (1999). "A Co-Evolutionary Approach to Playing Checkers Using Only Win, Lose, or Draw," In *Symposium on Computational Intelligence II, Aerosense99,* edited by K. Priddy, P. Keller, J. Bezdek, and D. B. Fogel, Bellingham, WA: SPIE, in press.

Clark, D. (1997). "Deep Thoughts on Deep Blue," *IEEE Expert,* Vol. 12:4, p. 31.

Conrad, M. (1988). "Prolegomena to Evolutionary Programming." In *Advances in Cognitive Science: Steps toward Convergence,* edited by M. Kochen and H. M. Hastings. AAAS, pp. 150–168.

Fogel, D. B. (1991). "The Evolution of Intelligent Decision Making in Gaming," *Cybernetics and Systems,* Vol. 22, pp. 223–236.

Fogel, D. B. (1992). "Evolving Artificial Intelligence." Ph.D. diss., UCSD.

Fogel, D. B. (1993a). "Evolving Behaviors in the Iterated Prisoner's Dilemma," *Evolutionary Computation,* Vol. 1:1, pp. 77–97.

Fogel, D. B. (1993b). "Using Evolutionary Programming to Create Neural Networks That Are Capable of Playing Tic-Tac-Toe." In *Proc. IEEE Intern. Conf. on Neural Networks, 1993,* Piscataway, NJ: IEEE Press, pp. 875–880.

Fogel, D. B. (1994). "Evolutionary Programming: An Introduction and Some Current Directions," *Statistics and Computing,* Vol. 4, pp. 113–129.

Fogel, D. B., L. J. Fogel, W. Atmar, and G. B. Fogel (1992). "Hierarchic Methods of Evolutionary Programming." In *Proc. of the First Ann. Conf. on Evolutionary Programming,* edited by D. B. Fogel and W. Atmar. La Jolla, CA: Evolutionary Programming Society, pp. 175–182.

Fogel, L. J. (1964). "On the Organization of Intellect." Ph.D. diss., UCLA.

Fogel, L. J., A. J. Owens, and M. J. Walsh (1966). *Artificial Intelligence through Simulated Evolution.* New York: John Wiley.

Griffith, N.J.L., and M. Lynch (1997). "NeuroDraughts: The Role of Representation, Search, Training Regime and Architecture in a TD Draughts Player," Unpublished technical report, University of Limmerick, Ireland.

Hofstadter, D. R. (1985). *Metamagical Themas: Questing for the Essence of Mind and Pattern.* New York: Basic Books.

Kaindl, H. (1990). "Tree Searching Algorithms," In *Computers, Chess, and Cognition,* edited by T. A. Marsland and J. Schaeffer, New York: Springer, pp. 133–168.

Koza, J. R. (1992). *Genetic Programming.* Cambridge, MA: MIT Press.

Ljung, L. (1987). *System Identification: Theory for the User.* Englewood Cliffs, NJ: Prentice Hall.

Mallows, C. L. (1973). "Some Comments on C_p," *Technometrics,* Vol. 15, pp. 661–675.

Maynard Smith, J. (1982). *Evolution and the Theory of Games.* Cambridge: Cambridge University Press.

Mayr, F. (1988). *Toward a New Philosophy of Biology. Observations of an Evolutionist.* Cambridge, MA: Belknap Press.

Pierce, J. R. (1980). *An Introduction to Information Theory: Symbols, Signals, and Noise,* 2nd ed., New York: Dover.

Rapoport, A. (1966). "Optimal Policies for the Prisoner's Dilemma." Technical Report No. 50, Psychometric Laboratory, University of North Carolina, NIH Grant, MH-10006.

Reed, J., R. Toombs, and N. A. Barricelli (1967). "Simulation of Biological Evolution and Machine Learning," *J. Theoretical Biology,* Vol. 17, pp. 319–342.

Risannen, J. (1978). "Modeling by Shortest Data Description," *Automatica,* Vol. 14, pp. 465–471.

Samuel, A. L. (1959). "Some Studies in Machine Learning Using the Game of Checkers," *IBM J. of Res. and Dev.,* Vol. 3:3, pp. 210–219.

Schaeffer, J. (1996). *One Jump Ahead: Challenging Human Supremacy in Checkers,* Berlin: Springer.

Schaeffer, J., R. Lake, P. Lu, and M. Bryant (1996). "Chinook: The World Man–Machine Checkers Champion," *AI Magazine,* Vol. 17:1, pp. 21–29.

Schwarz, G. (1977). "Estimating the Dimension of a Model," *Ann. Statist.,* Vol. 6:2, pp. 461–464.

Scodel, A., J. S. Minas, P. Ratoosh, and M. Lipetz (1959). "Some Descriptive Aspects of Two-Person Non-Zero Sum Games," *J. Conflict Resolution,* Vol. 3, pp. 114–119.

Turing, A. M. (1950). "Computing Machinery and Intelligence," *Mind,* Vol. 59, pp. 433–460.

Wallace, C. S., and D. M. Boulton (1968). "An Information Measure for Classification," *Computer Journal,* Vol. 11:2, pp. 185–194.

Wieland, A. P. (1991). "Evolving Controls for Unstable Systems." In *Connectionist Models: Proc. of the 1990 Summer School,* edited by D. S. Touretsky, J. L. Elman, T. J. Sejnowski, and G. E. Hinton. San Mateo, CA: Morgan Kaufmann, pp. 91–102.

CHAPTER 6

PERSPECTIVE

6.1 EVOLUTION AS A UNIFYING PRINCIPLE OF INTELLIGENCE

The theory of evolution is the great unifying theory of biology. But the impact of the theory extends even further: Evolution serves as a unifying description of all intelligent processes. Whether learning is accomplished by a species, an individual, or a social group, every intelligent system adopts a functionally equivalent process of reproduction, variation, competition, and selection. Each such system possesses a unit of mutability for generating new behaviors and a reservoir for storing knowledge (i.e., a memory; cf. Davis, 1991, p. 3). Learning is accomplished through some form of random search and the retention of those "ideas" that provide the learning system with the greatest understanding of its environment (Atmar, 1976). The learning system evolves, adapting its behavior to achieve its goals in a range of environments.

For example, sociogenetic learning has developed in two orders of Insecta (Hymenoptera: ants, wasps, and bees, and Isoptera: termites) (Wilson, 1971). Although the origin of Isoptera is distinctly different from Hymenoptera, both have evolved a set of communicative mechanisms and caste behaviors that are quite similar. Chemical patterns are used to transmit information regarding trail following, food marking, alarm, assembly, recruitment, recognition, and so forth. Foraging appears to occur by random exploration (Wilson, 1971). An odor trail laid down by an ant can be described as an "idea" (Atmar, 1976). Ideas are reinforced through repeated traveling and remarking. Good ideas (those that increase the probability of survival of the colony) are maintained, amplified, and become the knowledge of the colony. Poor ideas are quickly forgotten; the pheromones evaporate.

Phylogenetic learning (arising within the line of descent) is the earliest form of intelligence and has given rise to ontogenetic (arising within the individual) and sociogenetic (arising within the group) learning mechanisms that serve to increase mutability and longevity substantially (Atmar, 1976). But the specific form or mechanism of learning adopted is irrelevant. Only the functional behavior of the system is exposed to selection. Species, individuals, and social groups rely on the same underlying physics to generate adaptive behavior despite using various genetic, neuronal, or cultural mechanisms, just as birds, mammals, and fish rely on the same physics for achieving flight even though the specific mechanisms employed are vastly different. Every system that incorporates the evolutionary processes of reproduction, variation, competition, and selection, by whatever means, inherently evokes inductive learning. Intelligence is not the end result of evolution; rather, it is woven into the process itself. Intelligence and evolution cannot be separated.

6.2   PREDICTION AND THE LANGUAGELIKE NATURE OF INTELLIGENCE

Behavioral error is the sole quality optimized by selection, and that behavior is judged solely on its merits of predicting the forthcoming environmental conditions correctly (Atmar, 1994). Effective prediction requires the association of remembered events that are correlated in time, instances where one or more events are harbingers for some future occurrence. For example, sailors observing swallows flying low learned that this behavior was often associated with an oncoming storm. This relationship has been "used to predict bad weather: When you see the swallows flying low, prepare for bad weather. 'Swallows flying low' often precedes 'a storm' " (Lovejoy, 1975, pp. 9–10). No causality is implied in this observation, only correlation. Low-flying swallows do not cause the storm. Intelligent behavior is based on appropriately associating observations (symbols) so that future events can be anticipated and suitable actions may be taken in light of expected circumstances.

The mechanism of repeatedly mutating, testing, and selecting predictive organisms based on their survival produces systems that recognize the associations between environmental symbols and that adapt their behavior in light of these correlations. Kutas and Hillyard (1980) demonstrated that substantial neurophysiological potentials are generated in the speech regions of the human neocortex if a string of words is terminated with a grammatically proper but nonsensical terminating word, in contrast to a logical word. It must be presumed that the brain operates in a constant state of prediction (Atmar, personal communication). The order of observed symbols implies the rules of the environment; that is, the environment takes on a languagelike characteristic. The appropriate prediction of future environmental symbols demonstrates an understanding of that language.

The languagelike nature of intelligence becomes more obvious when language itself is studied. Bennett (1977) indicated that it is possible to randomly generate text that appears very much like English simply by taking into account the higher-order transitions across symbols (i.e., letters, spaces, and apostrophes). In examining Act III of Shakespeare's *Hamlet,* Bennett calculated the relative frequency of all symbols (first-order) and then randomly generated a sequence of text:

> AOOAAORH ONNNDGELC TEFSISO VTALIDMA POESDHEMHIES-
> WON PJTOMJ ITL FIM TAOFERLMT O NORDEERH HMFIOMRETW
> OVRCA OSRIE IEOBOTOGIM NUDSEEWU WHHS AWUA HIDNEVE
> NL SELTS ILY L CPATSDM EOLF OSTOM. . . .

But when he assessed and utilized the transition frequencies at the fourth order, he generated

> A GO THIS BABE AND JUDGMENT OF TIMEDIOUS RETCH AND NOT
> LORD WHAT IF THE EASELVES AND DO AND MAKE AND BASE
> GATHEM I AY BEATELLOUS WE PLAY MEANS HOLY FOOL MOUR
> WORK FORM INMOOT DED DE COMOULD HAVE MANY JUDG-
> MENT. . . .

Hayes (1983) extended Bennett's procedure to the seventh- and eighth-order transitions and created the following:

> We will forthwith dispatch, and most sovereign reason, like a gulf doth hourly
> grow out of the undiscovered countries different honesty should o'er which now
> goes to heavens to wash it white as snow, thought, 'tis seen what a noble and feed
> upon whose weal depends and countries different with strings of disprized love.
> How now, Ophelia? Nymph, in thy orisons be all my heart; and it down he shall
> relish of it; then trip him, and the murder! Pray can translate beauty be the op-
> pressor's wrong, the insolence of his likeness; this fear it is a massy whell, fix'd on
> the oppressor's wrong, the undiscover'd countries different with your loneliness.
> We shall keep those arrant she'll tax him home; and, as I think you; I your lord-
> ship?

and

> Did he receive them. No, not I; I never gave you well, well, well. Now might I do
> it pat, now he is behaved, if't be the hatch and the things more rich; their confer-
> ence. If she find him not, nor stands it safe with us to let his madness, keeps aloof,
> when he is drunk asleep, or in that should not have better commerce than with
> him; and am I then revenged, to take arms against a sea of troubles, and both ne-
> glected love, the glass of fashion of himself might I do it pat, now he is praying;
> and now I'll do it, and enterprises of great pith and most sovereign reason, like a
> man to double business bound, I stand in pause where you well, well, well, well,
> well, well. Now might I do it pat, now he is fit and sweat under a weary life, but
> like a man to double business bound, I stand in pause when I shall relish of sal-
> vation in't; then trip him, you sweet heavens! If thou dost marry, marry a fool; for
> which I did the murder?

Human response to even fourth- or fifth-order approximation of English "demonstrates the peculiar human compulsion to find pattern and meaning even where there is none" (Hayes, 1983). We are so conditioned to be predictive organisms that we are compelled to invent regularities in observed data, even when faced with purely random gibberish.

The English language may be regarded as somewhat complex. Yet an environment as complex as English text can be reasonably well predicted using no more than seventh- or eighth-order transitions. This has implications on the nature of creative genius. Bennett (1977) noted that "Mozart published a pamphlet explaining how to compose 'as many German Waltzes as one pleases' by throwing dice (see Scholes, 1950, plate 37)." Bennett (1977) speculated that artistic genius may be a "simple" matter of having the "requisite high-order correlation data," with the result being a random choice with an appropriate weighting as indicated above. "Similar conjecture could be made about the scientific thought-process as well. . . . Even in science the initial creative thought-process frequently arises from some sort of free-associative daydreaming, which is probably equivalent in a sense to repeatedly dragging out a bunch of correlation matrices" (Bennett, 1977).

Intelligent behaviors often appear languagelike because prediction is both the essence of intelligence and is essential to the use of language. A language is defined by its symbols and their associated regularities, the rules that define the transitions across symbols. A continuously generated sequence of symbols, in any language, conveys the least measure of information (in the Shannon sense of unexpected variation) when it is completely predictable. Under such conditions, the observed sequence generates no surprise. The physical environment, essentially, presents evolving species with a continuous sequence of symbols from a language. Species' behaviors evolve to reflect the communication from the environment. Selection operates to minimize species' behavioral surprise—that is, predictive error—as they are exposed to successive symbols. Behavioral surprise can never be brought to zero, or any absolute minimum, however, because the environment always contains some element of pure randomness and because the environment is itself affected by selection acting on the species that predict it.

Prediction was implicit in each experiment presented in Chapter 5. Each move in the iterated prisoner's dilemma elicited a sequence of actions and reactions from both players, and the surviving parents were ultimately good predictors of the behavior of their opponents. The appropriateness of any move in the games of tic-tac-toe or checkers depended on the entire anticipated sequence of moves that was to follow. Placing a marker to block an opponent's possible win in tic-tac-toe was only of benefit if the opponent would in fact take advantage of the presented opportunity. Setting up double jumps in checkers was only useful if the opponent's counter moves could be anticipated. In each case, the evolutionary simulation learned to predict its surroundings effectively. It learned the implicit language of its environment.

6.3 THE MISPLACED EMPHASIS ON EMULATING GENETIC MECHANISMS

The primary evolutionary force that shapes adaptive characteristics is natural selection (Wilson, 1992, p. 75). Selection operates not on individual genes, but rather on the entire holistic phenotype that is determined by the complete genetic complement of an organism and its interaction with its environment (Spetner, 1964; and many others). In this light, models of evolution that celebrate genetic detail or rely on an operational definition of evolution as simply a change in gene frequencies have been criticized by Mayr (1963, 1982, 1988), Hoffman (1989), Atmar (1994), and many others. Yet these reductionist approaches persist. Proponents of methods for simulating evolution that employ specific abstractions of genetic transfer mechanisms have offered a different view: "Evolution is a process that operates on chromosomes rather than on the living beings they encode. . . . The process of reproduction is the point at which evolution takes place" (Davis, 1991, p. 2).

But there is no single "point" at which evolution takes place. Evolution is a continuous process comprising four operations: reproduction, mutation, competition, and selection. To identify genetic reproduction or coding structures as the keystone to evolution is to misrepresent the process. Chromosomes, genes, codons, and nucleotides are not the beneficiaries of evolutionary optimization.

Holland (1975, p. 109) offered, "Though mutation is one of the most familiar of the genetic operators, its role in adaptation is frequently misinterpreted. In genetics, mutation is a process wherein one allele of a gene is randomly replaced by (or modified to) another to yield a new structure." Such a definition may be useful in the study of and classification of possible alterations to the genetic program. But such a definition is not very useful or meaningful in terms of mutation's "role in adaptation." A mutation is simply a change in genetic material (Hartl and Clark, 1989, p. 96) and the phenotypic effect (i.e., the role in adaptation) of a random mutation can almost never be predicted with any reasonable accuracy.

It is inappropriate to model evolution as a simple process of genetic mechanisms because these mechanisms are unimportant except for the behaviors that they portend. "It is less important for the understanding of evolution to know how genetic variation is brought about than to know how natural selection deals with it" (Mayr, 1970, p. 8). The ubiquity of convergent evolution, in which strikingly similar phenotypic traits arise independently despite completely different morphological and genetic backgrounds, provides support for this perspective. Although determining the genetic bases for these phenotypic traits is of interest, such investigation may not provide a greater understanding of any trait's role in adaptation. The term *adapt* means literally *to fit,* and fitness is a property of the behavior generated by the genetic program, not directly of the program itself. The same genetic program can generate different phenotypes under varying environments, and any measure of appropriateness can only be attributed to the phenotype. Evolution is primarily a process of adaptive behavior, rather than adaptive genetics. The behavioral relationship between successive generations of organisms should be the paramount interest of abstraction, not the genetic relationship.

Simulations of evolution can be constructed to incorporate crossing-over, inversion, sexual recombination, jumping genes, introns, and other genetic attributes. But these are simply symptoms of the process of evolution, and attention should instead be devoted to pursuing a fundamental understanding of the underlying physics. For example, flapping wings are the most overt symptom of naturally evolved flight, and they are fundamental to flight in birds, but only as a mechanism for generating airflow over an airfoil. "No practical, man-carrying ornithoper has ever been built. . . . The secret to the physics of flight does not lie in the obvious flaps of the wings but in the far less obvious shape of the airfoil" (Atmar, 1992). No single symptom of genotypic transformation (e.g., crossing-over) is solely responsible for the observed patterns of behavior in the vast majority of complex systems. The overt simulation of symptoms is generally a misplaced effort.

6.4   BOTTOM-UP VERSUS TOP-DOWN

In general, there are two approaches to analyzing complex information structures. Bottom-up approaches segregate and emphasize individual components and their local interactions, decomposing the total system into successively smaller subsystems that are then analyzed in a piecemeal fashion. Top-down analyses emphasize extrinsically imposed forces and their associated physics (to the extent known). Neither approach is wholly satisfactory because each is a simplification of the true system, but neither do they possess equal explanatory power (Atmar, 1992).

There are inherent pitfalls in bottom-up analyses. As a system is decomposed into smaller and smaller subsystems, the interactions between these become increasingly difficult to observe until, in the limit, they are completely lost. In subsequent analysis of each tractable subsystem, there is a great propensity to attempt to optimize each component for its supposed task without regard to how it may interact with other components. Upon reassembling the entire system, all interactions become unpredictable and must be described in terms of emergent properties, patterns of behavior that could not be forecasted from analysis of isolated components. "The necessity to resort to emergent properties is a hallmark of all explanatory hypotheses which reverse the actual flow of causation" (Atmar, 1992).

Atkinson (1977, p. 6) cautioned against analyzing systems bottom-up: "An inexperienced designer often finds it necessary to overcome a tendency toward overdesign of components. He may lavish much loving care on the design of a truly superior variable-reluctance, temperature-compensated, self-orienting widget, but this perfection of one part may detract from the utility of the device in which it is to be used." But natural selection does not optimize structures from the bottom-up. Only the entire cohesive individual is evaluated through competition in its environment, and each behavioral component affects total system performance in many ways. For example, flight in birds offers clear advantages, but perhaps many more than one might first realize: increased range for foraging, increased environment for prey (including airborne prey), more effective stalking, greater chance for escape from predators, ability to nest eggs in locations that are inaccessible to many predators, and so forth (Atkinson, 1977, p. 7). As natural selection acts on the complete polyparametric phenotype, it is impossible for any single behavioral parameter to be optimized without regard to the manner in which it interacts with the phenotype as a whole.

In contrast, evolutionary algorithms are sometimes still described as using the bottom-up mechanism of putting together good "building blocks and eventually combining these to get larger building blocks" (Goldberg, 1983; also Goldberg, 1989, p. 45; Davis, 1991, p. 19; Holland, 1992, p. 180; Forrest, 1993; cf. Grefenstette, 1991, 1993). In this context, building blocks are analogous to functional subunits. But the identification of "good" building blocks implies a criterion by which to measure such genetic units, a method for assigning credit to them.

The credit assignment problem has been widely studied and discussed (Minsky, 1961; Doran, 1967; Holland, 1980, 1985; Cohen and Feigenbaum, 1982; Barto et al., 1983; Booker et al., 1990; and others). Minsky (1961) commented, "In playing a complex game such as chess or checkers, or in writing a computer program, one has a definite success criterion: the game is won or lost. But in the course of play, each ultimate success (or failure) is associated with a vast number of internal decisions. If the run is successful, how can we assign credit for the success among the multitude of decisions?" Newell (in Minsky, 1961) stated: "It is extremely doubtful whether there is enough information in 'win, lose, or draw' when referred to the whole play of the game to permit any learning at all over available time scales." But, metaphorically, "win, lose, or draw" is the only information on which natural selection operates.

There are no viable credit assignment algorithms for isolated genetic structures or behavioral traits, nor can there be in any system where behavioral traits are highly integrated and the relationship between the genotype and the phenotype is characterized by pleiotropy and polygeny. Credit assignment is a nonexistent problem in evolution. It is a human construction. Different algorithms for distributing credit have been offered and may be useful in certain applications. But such methods are not required for successful optimization and will generally lead to suboptimal (but perhaps adequate) performance. In contrast, a top-down approach that only emphasizes total system performance can discover optimal behavior in complex situations. The experiments in Chapter 5 with evolutionary algorithms achieved a high level of play in the iterated prisoner's dilemma, in tic-tac-toe, and in checkers without the use of credit assignment algorithms.[1]

Bottom-up analyses are also typically fraught with after-the-fact explanations. The geocentric model of the solar system is classical in this regard. After retrograde motion was observed and found inconsistent with the model, epicycles were added to the planetary orbits. "To explain the observations, Ptolemy had to assume that each superior planet revolved in its epicycle at just the right rate so that it reached perigee at the moment of opposition on every orbit" (Lightman and Gingerich, 1992). If enough epicycles upon epicycles were added, a sufficiently good fit to the data could be obtained. But such a model assumes epicycles as a "fact-in-itself" (Frisius, 1560) rather than a fact reasoned from the model. The heliocentric explanation offered by Copernicus infers retrograde motion, but explains it rather than merely accepts it.

Similar problems are associated with evolutionary simulations that are conducted from the bottom-up. When the observed evidence of evolution is in contradiction with the simulated results, there is a tendency to want to fix or improve the results by incorporating additional operations. For example, De Jong (1975) noted that the performance of a certain evolutionary algorithm could be improved by making it more difficult for particular schemata to dominate the population (see also Goldberg, 1989, p. 116). To avoid this premature convergence, De Jong (1975) incorporated a clever procedure motivated by

---

[1] Note that the coevolutionary checkers experiments in Chapter 5 demonstrate that Newell's conjecture was fundamentally not correct. Granted, if taken in the strictest, most literal interpretation, the potential to learn proficient play in complex games based only on "win, lose, or draw" is likely to require an infeasible amount of time. But this is an unreasonable interpretation, particularly because people do not begin to learn games such as checkers *tabula rasa*: They perceive patterns borrowed from common events in life, they bring forward heuristics from other games played, and so forth. Newell's assertion, when taken in the context it was offered, was quite specific: "For learning to take place, each play of the game *must* yield much more information. This is . . . achieved by breaking the problem into components. The unit of success is the goal. If a goal is achieved, its subgoals are reinforced; if not they are inhibited" [italics added for emphasis]. Minsky (1961) offered "We are in complete agreement with Newell on this approach to the problem." Although this reinforcement learning can be effective, Newell's (and Minsky's) speculation that it is *necessary* is disproved by counterexample (see Chapter 5).

the notion of overpopulation (De Jong, personal communication). New solutions were made to replace solutions that were most similar based on a bit-by-bit comparison (i.e., "crowding factors"). Even though this operation can accelerate optimization and even though it has a biological inspiration, it is not characteristic of natural evolution. Natural selection acts only on the phenotypic appropriateness of individuals, not on their genetic similarity.

The natural evidence itself indicates whether or not a system is being optimized from the bottom-up. If it is, then individual coding segments must become a collection of "perfected 'genes,' informationally disconnected from one another, each operating to maximize its own optimal survival" (Atmar, 1992). But if, on the other hand, the physics of selection operates top-down, where only total system performance is evaluated and the underlying code is never directly affected but only indirectly affected, the result will be quite different. Behavior will become highly integrated functional "organs of extreme perfection" (Darwin, 1859, ch. 6), while the genetic basement will become a mélange of interwoven, overlain, and duplicated code. Only this latter pattern is observed in nature.

6.5 TOWARD A NEW PHILOSOPHY OF MACHINE INTELLIGENCE

Research in AI has typically passed over investigations into the primal causative factors of intelligence to more rapidly obtain the immediate consequences of intelligence (Atmar, 1976). Yet many of these efforts have not defined the intelligence they attempt to create. Efficient theorem proving, pattern recognition, and tree searching are symptoms of intelligent behavior. But systems that can accomplish these feats are not, simply by consequence, intelligent. Without a working definition of what constitutes intelligence, the field of research has fragmented into a collection of methods and tricks for solving particular problems in restricted domains of interest.

Certainly, these methods have been successfully applied to specific problems. Knowledge-based chess-playing programs have achieved a level of play that is competitive with or even superior to human masters. Fuzzy systems have been applied to image processing, industrial control, and robotics. Neural networks have been used for various signal processing, spatio-temporal pattern recognition, and automatic control applications, among others. Hybrids of these methods have been applied to a great range of problems (e.g., Gallant, 1988; Kandel, 1991; Kosko, 1992). But most of these applications require human expertise. They may be impressively applied to difficult problems, but they generally do not advance our understanding of intelligence. They solve problems, but they do not solve the problem of how to solve problems.

Evolution provides the solution to the problem of how to solve problems.

Moreover, the greatest gain from evolutionary computation will come when these algorithms are used to address problems that have, to date, been resistant to solution by human expertise, for evolutionary computation is not limited to recapitulating human behavior. "Recognizing that the scientific method is an evolutionary process that can be simulated opens the possibility of automating higher levels of intellect" (Fogel et al., 1966, p. 123). Such computational efforts can usefully combine research in expert, fuzzy, and connectionist systems. But any real advancement of evolutionary computation should rely on the careful observation and abstraction of the natural process of evolution.

The majority of efforts in artificial intelligence have simulated specific facets of intelligent behavior, mainly as we observe them in ourselves. Yet intelligence cannot be restricted to humans. Indeed, the life process itself provides the most common form of intelligence (Atmar, 1976). Darwinian evolution accounts for the behavior of naturally occurring intelligent entities and can be used to guide the creation of artificial entities that are capable of intelligent behavior. But more than that, if the evolutionary approach to artificial intelligence is adopted, there will truly be nothing "artificial" about it. Evolutionary machines will simply be intelligent, and the limit of such intelligence is unknowable.

REFERENCES

Atkinson, D. E. (1977). *Cellular Energy Metabolism and Its Regulation.* New York: Academic Press.

Atmar, J. W. (1976). "Speculation on the Evolution of Intelligence and Its Possible Realization in Machine Form." Sc.D. diss., New Mexico State University, Las Cruces.

Atmar, W. (1986). Personal Communication, AICS Research, Inc., Las Cruces, NM.

Atmar, W. (1992). "The Philosophical Errors That Plague both Evolutionary Theory and Simulated Evolutionary Programming." In *Proc. of First Ann. Conf. on Evolutionary Programming,* edited by D. B. Fogel and W. Atmar. La Jolla, CA: Evolutionary Programming Society, pp. 27–32.

Atmar, W. (1994). "Notes on the Simulation of Evolution," *IEEE Trans. Neural Networks,* Vol. 5:1, pp. 130–148.

Barto, A. G., R. S. Sutton, and C. W. Anderson (1983). "Neuronlike Adaptive Elements That Can Solve Difficult Learning Control Problems," *IEEE Trans. Sys., Man, and Cybern.,* Vol. SMC-13:5, pp. 834–846.

Bennett, W. R. (1977). "How Artificial Is Intelligence?" *American Scientist,* November–December, pp. 694–702.

Booker, L. B., D. E. Goldberg, and J. H. Holland (1990). "Classifier Systems and Genetic Algorithms." In *Machine Learning: Paradigms and Methods,* edited by J. Carbonell. Cambridge, MA: MIT Press, pp. 235–282.

Cohen, P. R., and E. A. Feigenbaum (1982). *The Handbook of Artificial Intelligence,* Vol. 3. Los Altos, CA: Morgan Kaufmann.

Darwin, C. (1859). *On the Origin of Species by Means of Natural Selection or the Preservations of Favored Races in the Struggle for Life.* London: John Murray.

Davis, L., ed. (1991). *Handbook of Genetic Algorithms.* New York: Van Nostrand Reinhold.

De Jong, K. A. (1975). "The Analysis of the Behavior of a Class of Genetic Adaptive Systems." Ph.D. diss., University of Michigan, Ann Arbor.

De Jong, K. A. (1994). Personal Communication, George Mason University.

Doran, J. (1967). "An Approach to Automatic Problem Solving." In *Machine Learning,* Vol. 1, edited by E. L. Collins and D. Michie. Edinburgh: Oliver and Boyd, pp. 105–123.

Fogel, L. J., A. J. Owens, and M. J. Walsh (1966). *Artificial Intelligence through Simulated Evolution.* New York: John Wiley.

Forrest, S. (1993). "Genetic Algorithms: Principles of Natural Selection Applied to Computation." *Science,* Vol. 261, pp. 872–878.

Gallant, S. (1988). "Connectionist Expert Systems," *Comm. of the ACM,* Vol. 31, pp. 152–169.

Frisius Reiner Gemma (1560). In Johannes Stadius, Ephemerides Novae et Auctae, Cologne, si, b3–b3cv, translated by O. Gingerich and R. S. Westman, *The Wittich Connection: Conflict and Priority in Late Sixteenth Century Cosmology,* Trans. Am. Philos. Soc., Vol. 78, part 7, p. 42, 1988.

Grefenstette, J. J. (1991). "Conditions for Implicit Parallelism," In *Foundations of Genetic Algorithms,* edited by G.J.E. Rawlins. San Mateo, CA: Morgan Kaufmann, pp. 252–261.

Grefenstette, J. J. (1993). "Deception Considered Harmful," In *Foundations of Genetic Algorithms 2,* edited by L. D. Whitley. San Mateo, CA: Morgan Kaufmann, pp. 75–91.

Goldberg, D. E. (1983). "Computer-Aided Gas Pipeline Operation Using Genetic Algorithms and Rule Learning," Ph.D. diss., University of Michigan, Ann Arbor.

Goldberg, D. E. (1989). *Genetic Algorithms in Search, Optimization, and Machine Learning.* Reading, MA: Addison-Wesley.

Hartl, D. L., and A. G. Clark (1989). *Principles of Population Genetics,* 2nd ed. Sunderland, MA: Sinauer.

Hayes, B. (1983). "A Progress Report on the Fine Art of Turning Literature into Drivel," *Scientific American,* Vol. 249:5, pp. 18–28.

Hoffman, A. (1989). *Arguments on Evolution: A Paleontologist's Perspective.* New York: Oxford University Press.

Holland, J. H. (1975). *Adaptation in Natural and Artificial Systems.* Ann Arbor: University of Michigan Press.

Holland, J. H. (1980). "Adaptive Algorithms for Discovering and Using General Patterns in Growing Knowledge-Bases," *Intern. J. Policy Anal. Inform. Syst.,* Vol. 4, pp. 217–240.

Holland, J. H. (1985). "Properties of the Bucket-Brigade." In *Proc. of an Intern. Conf. on Genetic Algorithms and Their Applications,* edited by J. J. Grefenstette. Pittsburgh, PA: Carnegie Mellon, pp. 1–7.

Holland, J. H. (1992). *Adaptation in Natural and Artificial Systems,* 2nd ed. Cambridge, MA: MIT Press.

Kandel, A. (1991). *Fuzzy Expert Systems*. Boca Raton, FL: CRC Press.

Kosko, B. (1992). *Neural Networks and Fuzzy Systems: A Dynamical Systems Approach to Machine Intelligence*. Englewood Cliffs, NJ: Prentice Hall.

Kutas, M., and S. A. Hillyard (1980). "Reading Senseless Sentences: Brain Potentials Reflect Semantic Incongruity," *Science,* Vol. 207, pp. 203–205.

Lightman, A., and O. Gingerich (1992). "When Do Anomalies Begin?" *Science,* Vol. 225, pp. 690–695.

Lovejoy, E. P. (1975). *Statistics for Math Haters*. New York: Harper and Row.

Mayr, E. (1963). *Animal Species and Evolution*. Cambridge, MA: Belknap Press.

Mayr, E. (1970). *Populations, Species, and Evolution*. Cambridge, MA: Belknap Press.

Mayr, E. (1982). *The Growth of Biological Thought: Diversity, Evolution, and Inheritance*. Cambridge, MA: Belknap Press.

Mayr, E. (1988). *Toward a New Philosophy of Biology: Observations of an Evolutionist*. Cambridge, MA: Belknap Press.

Minsky, M. L. (1961). "Steps toward Artificial Intelligence," *Proc. of IRE,* Vol. 49:1, pp. 8–30.

Scholes, P. A. (1950). *The Oxford Companion to Music*. New York: Oxford University Press.

Spetner, L. M. (1964). "Natural Selection: An Information-Transmission Mechanism for Evolution," *J. Theo. Biol.,* Vol. 7, pp. 412–429.

Wilson, E. O. (1971). *The Insect Societies*. Cambridge, MA: Belknap Press.

Wilson, E. O. (1992). *The Diversity of Life*. New York: W. W. Norton.

GLOSSARY

Adaptation.   The process of generating a set of behaviors that more closely match or predict a specific environmental regime. An increased ecological-physiological efficiency of an individual relative to others in the population.

Adaptive Topography.   A continuous function describing the fitness of individuals and populations.

Agnathans.   A jawless class of vertebrates represented today by lampreys and hagfishes.

Allele.   An alternative form of a gene that occurs at a given chromosomal site (locus).

Altruistic Behavior.   The aiding of another individual (or entity) at one's own risk or to one's own detriment.

Asexual Reproduction.   A form of reproduction that does not involve the use of gametes.

Behavior.   The response of an organism to the present stimulus and its present state. It is the total sum of behaviors of an organism that define the fitness of the organism to its present environment; thus it is the operative function against which selection operates.

Chiasmata.   The X-shaped, microscopically visible figures representing homologous chromatids that have exchanged genetic material through crossing-over during meiosis.

Chromatids.   The two identical parts of a duplicated chromosome.

Chromosome.   Rod-shaped bodies in the nucleus of eukaryotic cells, most visible particularly during cell division, which contain the hereditary units or genes.

Codon.   A sequence of three nucleotide bases contained within the structure of the DNA molecule (or mRNA) that codes for a specific amino acid or for a termination signal during protein synthesis.

Convergent Evolution. The independent development of similar structures in organisms that are not directly related.

Crossing-over. The exchange of corresponding segments of genetic material between chromatids of homologues at meiosis.

Diploid. The 2N number of chromosomes; the complete or total number; twice the number of chromosomes found in gametes.

DNA. Deoxyribonucleic acid.

Engram. Supposed pathway of learned neural connections.

Ethology. Patterns of animal behavior. The study of patterns of animal behavior.

Eukaryotic Cell. Cell with a membrane-enclosed nucleus and organelles found in animals, fungi, plants, and protists.

Fitness. A summation of the quality of environmental prediction by an organism throughout its range of regimes. The probability of or propensity for survival of an individual or population.

Founder effect. A cause of genetic drift attributable to colonization by a limited number of emigrants from a parent population.

Gamete. A reproductive cell that joins with another in fertilization to form a zygote, most often an egg or sperm.

Gene. A unit of heredity located on a chromosome and composed of DNA. The code necessary to promote the synthesis of one polypeptide chain.

Genetic Drift. Changes to the gene pool of a population by chance processes alone.

Genome. The total genetic constitution of an organism.

Genotype. The sum of inherited characters maintained within the entire reproducing population. Often also used to refer to the genetic constitution underlying a single trait or set of traits.

Haploid. The N number of chromosomes. See *diploid*.

Heredity. The transmission of characteristics from parent to offspring through the gametes.

Heterozygous. Having two different alleles for a given trait.

Homologues. Duplicated chromosomes that look alike and have genes affecting the same traits.

Homozygous. Having identical alleles for a given trait.

Intron. An intervening noncoding region of a eukaryotic gene.

Inversion. A reversal in order of a segment of a chromosome.

Linkage. Genes on the same chromosome are linked in the sense that they tend to move together to the same gamete. Crossing-over interferes with linkage.

Locus.   A particular location on a chromosome.

Meiosis.   Type of cell division that occurs during the production of gametes, by means of which the daughter cells receive the haploid number of chromosomes.

Morphology.   Form and its development. Also, the study of form and its development.

Mutation.   A genetic change.

Natural Selection.   The result of competitive exclusion as organisms fill the available finite resource space.

Neutral Mutation.   A mutation in the genotypic structure of a population that promotes no behavioral difference in the present environment.

Nucleus.   A large organelle containing the chromosomes and acting as a control center for the cell.

Ontogenetic.   Literally "arising within the self." As a learning process, learning arising within the individual.

Parthenogenesis.   The development of an egg without fertilization.

Phenotype.   The behavioral expression of the genotype in a specific environment. The realized expression of the genotype. The functional expression of a trait.

Pheromones.   Small, volatile chemical signals functioning in communication between animals and acting much like hormones to influence physiology and behavior.

Phylogenetic.   Literally, "arising within the line of descent." As a learning process, learning arising within the phyletic line of descent.

Phylogeny.   The evolutionary relationships among any group of organisms.

Pleiotropy.   The capacity of a gene to affect a number of different phenotypic characteristics.

Polygeny.   The circumstance where a single phenotypic trait is affected by multiple genes.

Prokaryotic Cell.   A cell lacking a membrane-enclosed nucleus and organelles.

Saltation.   An abrupt variation in the condition of a species; a marked modification.

Sociogenetic.   Literally, "arising within the group." As a learning process, learning arising within a social group.

Species.   A group of similarly constructed organisms that are capable of interbreeding and producing fertile offspring. A population whose members are able to interbreed freely under natural conditions.

Synapsis.   The attracting and pairing of homologous chromosomes during meiosis.

Teleonomic Behavior.   Behavior that owes its goal-directedness to the operation of a program.

Zygote.   Fertilized egg that is always diploid.

LITERATURE CITED FOR GLOSSARY

Atmar, J. W. (1976). "Speculation on the Evolution of Intelligence and Its Possible Realization in Machine Form." Sc.D. diss., New Mexico State University, Las Cruces.

Campbell, N. A. (1990). *Biology,* 2nd ed. Redwood City, CA: Benjamin/Cummings.

Hartl, D. L., and A. G. Clark (1989). *Principles of Population Genetics*, 2nd ed. Sunderland, MA: Sinauer.

Mader, S. S. (1983). *Inquiry into Life,* 3rd ed. Dubuque, IA: William C. Brown.

Mayr, E. (1988). *Toward a New Philosophy of Biology: Observations of an Evolutionist.* Cambridge, MA: Belknap Press.

Raven, P. H., and G. B. Johnson (1986). *Biology.* St. Louis: Times Mirror/Moseby College.

Wilson, E. O. (1992). *The Diversity of Life.* New York: W. W. Norton.

INDEX

ABOUT THE AUTHOR

David B. Fogel is executive vice president and chief scientist of Natural Selection, Inc. in La Jolla, CA — a small business focused on solving difficult problems in industry, medicine, and defense using evolutionary computation, neural networks, fuzzy systems, and other methods of computational intelligence. Dr. Fogel's experience in evolutionary computation spans 15 years and includes applications in pharmaceutical design, computer-assisted mammography, data mining, factory scheduling, financial forecasting, traffic flow optimization, agent-based adaptive combat systems, and many other areas. Prior to cofounding Natural Selection, Inc. in 1993, Dr. Fogel was a systems analyst at Titan Systems, Inc. (1984–1988), and a senior principal engineer at ORINCON Corporation (1988–1993).

Dr. Fogel received his Ph.D. degree in engineering sciences (systems science) from the University of California at San Diego (UCSD) in 1992. He earned an M.S. degree in engineering sciences (systems science) from UCSD in 1990, and a B.S. in mathematical sciences (probability and statistics) from the University of California at Santa Barbara in 1985. He has taught university courses at the graduate and undergraduate level in stochastic processes, probability and statistics, and evolutionary computation. Dr. Fogel is a prolific author in evolutionary computation, having published over 40 journal papers, as well as over 80 conference publications, 20 contributions in book chapters, one video, and three books—most recently, *Evolutionary Computation: The Fossil Record* (IEEE Press, 1998). In addition, Dr. Fogel is coeditor in chief of the *Handbook of Evolutionary Computation* (Oxford, 1997) and the founding editor-in-chief of the *IEEE Transactions on Evolutionary Computation*. He serves as associate editor for the journal *BioSystems* and is a member of the editorial board of several other international technical journals.

Dr. Fogel served as a Visiting Fellow of the Australian Defence Force Academy in November 1997, and is a member of many professional societies including the American Association for the Advancement of Science, the

American Association for Artificial Intelligence, Sigma Xi, and the New York Academy of Sciences. He was the founding president of the Evolutionary Programming Society in 1991 and is a Fellow of the IEEE, as well as an associate member of the Center for the Study of Evolution and the Origin of Life (CSEOL) at the University of California at Los Angeles. Dr. Fogel is a frequently invited lecturer at international conferences and a guest for television and radio broadcasts.